Cheney 1989 April

SPARSE MATRICES:
Numerical Aspects with Applications for Scientists and Engineers

MATHEMATICS AND ITS APPLICATIONS
Series Editor: G. M. BELL, Professor of Mathematics,
King's College London (KQC), University of London

NUMERICAL ANALYSIS, STATISTICS AND OPERATIONAL RESEARCH
Editor: B. W. CONOLLY, Emeritus Professor of Mathematics (Operational Research),
Queen Mary College, University of London

Mathematics and its applications are now awe-inspiring in their scope, variety and depth. Not only is there rapid growth in pure mathematics and its applications to the traditional fields of the physical sciences, engineering and statistics, but new fields of application are emerging in biology, ecology and social organization. The user of mathematics must assimilate subtle new techniques and also learn to handle the great power of the computer efficiently and economically.

The need for clear, concise and authoritative texts is thus greater than ever and our series will endeavour to supply this need. It aims to be comprehensive and yet flexible. Works surveying recent research will introduce new areas and up-to-date mathematical methods. Undergraduate texts on established topics will stimulate student interest by including applications relevant at the present day. The series will also include selected volumes of lecture notes which will enable certain important topics to be presented earlier than would otherwise be possible.

In all these ways it is hoped to render a valuable service to those who learn, teach, develop and use mathematics.

Mathematics and its Applications
Series Editor: G. M. BELL, Professor of Mathematics, King's College London (KQC), University of London

Anderson, I.	Combinatorial Designs
Artmann, B.	Concept of Number: From Quaternions to Monads and Topological Fields
Arczewski, K. & Pietrucha, J.	Mathematical Modelling in Discrete Mechanical Systems
Arczewski, K. and Pietrucha, J.	Mathematical Modelling in Continuous Mechanical Systems
Bainov, D.D. & Konstantinov, M.	The Averaging Method and its Applications
Baker, A.C. & Porteous, H.L.	Linear Algebra and Differential Equations
Balcerzyk, S. & Jösefiak, T.	Commutative Rings
Balcerzyk, S. & Jösefiak, T.	Commutative Noetherian and Krull Rings
Baldock, G.R. & Bridgeman, T.	Mathematical Theory of Wave Motion
Ball, M.A.	Mathematics in the Social and Life Sciences: Theories, Models and Methods
de Barra, G.	Measure Theory and Integration
Bartak, J., Herrmann, L., Lovicar, V. & Vejvoda, D.	Partial Differential Equations of Evolution
Bell, G.M. and Lavis, D.A.	Co-operative Phenomena in Lattice Models, Vols. I & II
Berkshire, F.H.	Mountain and Lee Waves
Berry, J.S., Burghes, D.N., Huntley, I.D., James, D.J.G. & Moscardini, A.O.	Mathematical Modelling Courses
Berry, J.S., Burghes, D.N., Huntley, I.D., James, D.J.G. & Moscardini, A.O.	Mathematical Modelling Methodology, Models and Micros
Berry, J.S., Burghes, D.N., Huntley, I.D., James, D.J.G. & Moscardini, A.O.	Teaching and Applying Mathematical Modelling
Blum, W.	Applications and Modelling in Learning and Teaching Mathematics
Brown, R.	Topology: A Geometric Account of General Topology, Homotopy Types and the Fundamental Groupoid
Burghes, D.N. & Borrie, M.	Modelling with Differential Equations
Burghes, D.N. & Downs, A.M.	Modern Introduction to Classical Mechanics and Control
Burghes, D.N. & Graham, A.	Introduction to Control Theory, including Optimal Control
Burghes, D.N., Huntley, I. & McDonald, J.	Applying Mathematics
Burghes, D.N. & Wood, A.D.	Mathematical Models in the Social, Management and Life Sciences
Butkovskiy, A.G.	Green's Functions and Transfer Functions Handbook
Cartwright, M.	Fourier Methods: Applications in Mathematics, Engineering and Science
Cerny, I.	Complex Domain Analysis
Chorlton, F.	Textbook of Dynamics, 2nd Edition
Chorlton, F.	Vector and Tensor Methods
Cohen, D.E.	Computability and Logic
Cordier, J.-M. & Porter, T.	Shape Theory: Categorical Methods of Approximation
Crapper, G.D.	Introduction to Water Waves
Cross, M. & Moscardini, A.O.	Learning the Art of Mathematical Modelling
Cullen, M.R.	Linear Models in Biology
Dunning-Davies, J.	Mathematical Methods for Mathematicians, Physical Scientists and Engineers
Eason, G., Coles, C.W. & Gettinby, G.	Mathematics and Statistics for the Biosciences
El Jai, A. & Pritchard, A.J.	Sensors and Controls in the Analysis of Distributed Systems
Exton, H.	Multiple Hypergeometric Functions and Applications
Exton, H.	Handbook of Hypergeometric Integrals
Exton, H.	q-Hypergeometric Functions and Applications
Faux, I.D. & Pratt, M.J.	Computational Geometry for Design and Manufacture
Firby, P.A. & Gardiner, C.F.	Surface Topology
Gardiner, C.F.	Modern Algebra

Series continued at back of book

SPARSE MATRICES:
Numerical Aspects with Applications for Scientists and Engineers

U. SCHENDEL, Diph-Math., Dr.rernat.
Professor of Numerical Mathematics and Computer Science
Institut für Mathematik III, Freie Universität, Berlin
Federal Republic of Germany

Translation Editor:
BRIAN CONOLLY
Emeritus Professor of Operational Research
Queen Mary College, University of London

ELLIS HORWOOD LIMITED
Publishers · Chichester

Halsted Press: a division of
JOHN WILEY & SONS
New York · Chichester · Brisbane · Toronto

This English edition first published in 1989 by
ELLIS HORWOOD LIMITED
Market Cross House, Cooper Street,
Chichester, West Sussex, PO19 1EB, England
The publisher's colophon is reproduced from James Gillison's drawing of the ancient Market Cross, Chichester.

Distributors:

Australia and New Zealand:
JACARANDA WILEY LIMITED
GPO Box 859, Brisbane, Queensland 4001, Australia

Canada:
JOHN WILEY & SONS CANADA LIMITED
22 Worcester Road, Rexdale, Ontario, Canada

Europe and Africa:
JOHN WILEY & SONS LIMITED
Baffins Lane, Chichester, West Sussex, England

North and South America and the rest of the world:
Halsted Press: a division of
JOHN WILEY & SONS
605 Third Avenue, New York, NY 10158, USA

This English edition is translated from the original German edition *Sparse matrizen*, published in 1977 by R. Oldenbourg, Munich, © the copyright holders.

© 1989 English Edition, Ellis Horwood Limited

British Library Cataloguing in Publication Data
Schendel, U. (Udo)
Sparse matrices: numerical aspects with applications for scientists and engineers.
1. Algebra. Sparse matrices. Applications
I. Title II. Series
512.9′434
Library of Congress CIP data available

ISBN 0–7458–0635–X (Ellis Horwood Limited)
ISBN 0–470–21406–6 (Halsted Press)

Printed in Great Britain by Hartnolls, Bodmin

COPYRIGHT NOTICE
All Rights Reserved. No part of this publication may be reproduced, stored in a retrieval system, or transmitted, in any form or by any means, electronic, mechanical, photocopying, recording or otherwise, without the permission of Ellis Horwood Limited, Market Cross House, Cooper Street, Chichester, West Sussex, England.

Table of contents

Preface		vii
1	**Introduction**	1
2	**Storage techniques**	5
	2.1 Linked lists	6
	2.2 Storage technique in case of identical matrix elements	7
	2.3 Unlinked lists	8
3	**Systems of linear equations**	11
	3.1 Introductory example	11
	3.2 Direct methods	12
	3.3 Analysis of the elimination process	14
	3.4 Organization of storage	18
	3.5 Minimizing the fill-in	20
	3.6 Stability considerations	22
4	**Computation of the inverse**	25
5	**Block matrices**	28
	5.1 Block elimination of linear sparse systems	28
	5.2 Property $\mathcal{P}$	31
	5.3 Direct algorithms with block matrices	35
	5.4 Least squares problems	41

6 Iterative algorithms 47
   6.1 Some properties of sparse matrices 48
   6.2 The Jacobi method 52
   6.3 The Gauss-Seidel method 57
   6.4 The SOR method 61
   6.5 The SOR method with block matrices 64
   6.6 Convergence criteria for SOR methods 66

7 Graphs and matrices 72

8 Eigenvalues and eigenvectors 79

9 Parallel numerical algorithms 87
   9.1 Basic concepts for the development of parallel algorithms 89
   9.2 Arithmetical expressions 92
   9.3 Some remarks on the development of parallel algorithms 95
   9.4 Recurrence relations 97
   9.5 Parallel solution of linear systems on SIMD machines 101
   9.6 Parallel solution of linear systems on MIMD machines 104

References 107

Index 112

Preface

This book has developed from lectures held at the Free University and Technical University of Berlin. Numerous talks with engineers have shown that there is a need to impart knowledge on large, sparse matrices and the mathematical methods which use these matrices. In this connection it is of great importance that it is only the combined effect of mathematical methods and digital computers which produces the effect of the chosen algorithm. The present book has been written with the intention to fill a gap by dicussing this relationship by means of some methods and by supporting it with examples.

In the selection of material the emphasis was put on numerical methods which belong to the classical, current procedures of numerical mathematics and which have already proved their efficiency in practical use; we did not aim at completeness of all possible procedures. The aim was rather to stress the typical features of the described methods and to achieve a better comprehension of the questions arising in connection with sparse matrices.

A mathematical representation has been chosen which will enable natural scientists, engineers and students of the natural sciences to find access to these problems. Knowledge in the programming in a higher computer language and basic knowledge in numerical mathematics are prerequisite. In particular, this book has been written for readers who work outside universities in practical fields and who want to get a general idea of the topic of sparse matrices.

Large parts of this book have already been published by R. Oldenbourg Publishers, Munich-Vienna.

The idea to translate the German version into English has led to a revision and an extension of the contents. The field of sparse matrix problems has received many a fresh impetus in the last few years.

My colleague, Mr. J. Brandenburger, has contributed essentially in revising and translating the German version of this book. He also produced the camera-ready copy by the word processing system LaTeX. The translation itself is mainly due to Mr. W. Pourie. I want to express to them my particular gratitude for their good cooperation.

Besides, I want to express my thanks to my colleague, Mr. B. Conolly, Queen Mary College, London, for the pains he took in critically reading and polishing the English translation.

Finally I want to thank the Ellis Horwood Publishing Company, London, for their good and sympathetic cooperation.

Berlin, August 1988   U. Schendel

1 Introduction

For about 20 years the different aspects of sparse matrices have been receiving increasing attention. Several important international conferences have taken place on sparse matrices and their applications with a wide range of topics.

In addition, there have been a number of smaller conferences about specific problems of sparse matrices like QR-methods, structural analysis, power distribution systems, circuit design and others. In numerical analysis most areas, but in particular eigenvalue problems, integral equations, linear and nonlinear equations, linear programming, ordinary and partial differential equations have been covered. In mathematics in a wide sense combinatorics, computing, graph theory and statistics are dealt with, too.

All these special problems lead to a matrix $A := [a_{ik}]$, $A \in I\!R^{m \times n}, \mathbb{C}^{m \times n}$, whose number r of elements a_{ik} with $a_{ik} \neq 0$ is small in relation to the total number $n \times m$ of all matrix elements.

Effectiveness of work with these matrices requires:

- special numerical algorithms that take account of the sparseness

- special storage techniques

- special programming techniques.

These requirements arise from the necessity for

- the results to be numerically acceptable

- the storage demands to be minimized because of the limited storage capacity

- computation time and computation costs to be minimized.

These criteria are of the greatest importance to the numerical analyst. In this context the question of the existence or uniqueness of the solution is quite often more easily answered than the fulfilment of the requirements listed above.

New computer generations allow larger problems to be solved. The VSLI-technology [1] makes it possible to build up highly efficient, economically-priced computers for solving special problems (for example, self-adjoint elliptic partial differential equations).

Examples

There is no exact definition of a sparse matrix. A $(n \times n)$ matrix $A := [a_{ik}]$ is said to be a *sparse matrix* if only a small percentage of all matrix elements a_{ik}, $i, k = 1, 2, \ldots, n$, is nonzero (in practice less than 10%).

For example, in circuit design the following structures can be found:

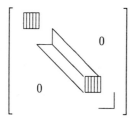

Other possible structures are:

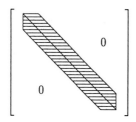   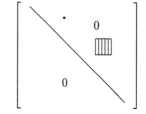

matrix with constant band width   arbitrary sparse matrix

---
[1] VSLI: Very Large Scale Integrated

1 Introduction

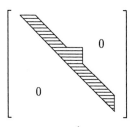

band matrix with step

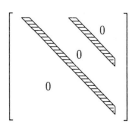

strip matrix

Condensed forms are desirable like:

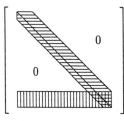

band matrix with margin

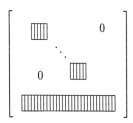

block diagonal matrix with margin

In general it has to be noted that every numerical problem involving sparse matrices must be treated individually; *adaptive algorithms* must be developed.

For example, in continuum mechanics the generalized form of the Poisson equation on a 2-dimensional domain $\mathcal{B}$

$$-\frac{\partial}{\partial x}(a \cdot \frac{\partial u}{\partial x}) - \frac{\partial}{\partial y}(c \cdot \frac{\partial u}{\partial y}) + k \cdot u = f, \quad (x,y) \in \mathcal{B} \subset \mathbb{R}^2 \qquad (1.1)$$

with $a = a(x,y) > 0$, $c = c(x,y) > 0$ and the boundary conditions

$$\alpha u + \frac{\partial u}{\partial n} = \beta, \quad (x,y) \in \Gamma \qquad (1.2)$$

has to be solved.

Γ : closed boundary of $\mathcal{B}$
n : direction of the normal derivation
α, β : piecewise continuous function on Γ.

Let $u = u(x,y)$ be a solution and k, f functions continuous within $\mathcal{B}$. In the method of finite elements $\mathcal{B}$ is partitioned into subdomains having forms such as triangles, rectangles and others, shown in Fig.1.1.

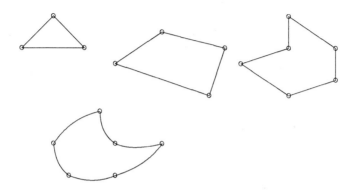

Fig. 1.1 – Examples for partitioned elements

The approximate solution $\hat{u}$ is a polynomial defined on each subdomain of $\mathcal{B}$ fulfilling certain boundary conditions. Finally triangularization of $\mathcal{B}$ leads to the problem of determining the minimal solution $\hat{u}$ of the algebraic equation

$$I(\hat{u}) = \frac{1}{2}(\hat{u}^T A \hat{u} - 2\hat{u}^T b), \qquad (1.3)$$

i.e. the system

$$\frac{\partial I}{\partial \hat{u}} = 0 \iff A\hat{u} = b \qquad (1.4)$$

has to be solved. The matrix A in this example is of large order, sparse, positive definite and symmetric.

2 Storage techniques

The effectiveness of work with sparse matrices depends not only on the underlying mathematics, but also on the extent to which the computer itself is an integral part of the algorithm. Thus storage techniques play an important role; their aim is to store information as densely and as economically as possible.

Example 2.1
In the frequency analysis of linear networks the linear systems $A(\omega_i)x = b$ have to be solved for different frequencies ω_i. A is Hermitian of order $n = 3304$ with 60685 nonzero elements; hence the density (i.e. the quotient of the number of nonzero elements and the total number of matrix elements) is 0.6 %. LU-factorization gives 105470 nonzero elements with a fill-in of 0.4 %. Conventional storage techniques would require about 20 million storage cells.

This example shows some of the properties which storage techniques must possess:

- only nonzero elements should be stored

- it should be possible to insert additionally created nonzero elements easily and quickly into the existing list of nonzero elements.

The treatment of sparse problems is affected considerably by the configuration of the computer available, i.e. according to their importance within the algorithms the data are stored either in fast-working and therefore expensive storages (for example core) or in slower peripheral storage (for example discs). At present various storage modules are available for users in medium-sized and larger equipment.

If an unstructured matrix is sufficiently sparse it can of course be kept throughout in the high-speed storage of a computer: only the elements $a_{ik} \neq 0$ are stored. The placement scheme for these elements in the store depends on the algorithm that needs these elements. Different kinds of storage are available. If the large matrix to be stored is of high density, or the number of nonzero elements exceeds the capacity of the core or high-speed storage, the matrix has to be kept in the low-speed external store and the matrix elements have to be transported in blocks into the high-speed storage. In this context the reader is recommended to study carefully the analysis of paging strategies for the solution of linear systems.

2.1 Linked lists

Each element $a_{ik} \neq 0$ in the column k is stored as an *item* I with

$$I := (i, v, p) \tag{2.2}$$

and

i : row index
v : value of the element a_{ik}
p : address of the next element $a_{ik} \neq 0$ of column k.

The address p is zero, if a_{ik} is the last nonzero element in the column k. Then an item can be depicted as follows:

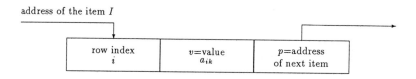

Fig. 2.1 – Item

Besides this block SI of store for the items a further block is needed to store the first address of each column:

BC : beginning of column address

Thus the total storage requirement S consists of both the part BC and part SI. Part BC has exactly n locations and SI requires exactly $3t$ storage locations, where t denotes the number of nonzero elements. The total length of S is: $n+3t$.

2.2 Storage technique in case of identical matrix elements

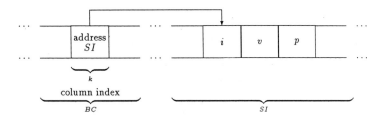

Fig. 2.2 – Total list

In consequence of this storage scheme additionally created nonzero elements can be stored in SI without having to rearrange them. To show how the creation of a new nonzero element affects BC and SI the following example is given:

Example 2.3

$$a_{13} := a_{33} = 0, \; a_{23} := 0.5, \; a_{43} := 1.5.$$

BS begins at location 101, and the items corresponding to a_{23} and a_{43} begin at locations 200 and 203 respectively; the element $a_{33} := 2.5$ is inserted later. This item starts at 300:

Table 2.1 Example

location	101	200	201	202	203	204	205	300	301	302
present contents	200	2	0.5	203	4	1.5	0	-	-	-
new contents	200	2	0.5	300	4	1.5	0	3	2.5	203
from		item $\hat{=} a_{23}$			item $\hat{=} a_{43}$			item $\hat{=} a_{33}$		
BS										

2.2 Storage technique in case of identical matrix elements

The values of the nonzero elements of a sparse matrix often are equal. In this case numerical constants are used. For illustration the following example is given:

Example 2.4

A $(5,5)$-sparse matrix A containing 13 nonzero elements is to be stored by columns.

$$A = \begin{bmatrix} 1.0 & 4.0 & 0 & 0 & 0 \\ 3.0 & 1.0 & 4.0 & 0 & 0 \\ 0 & 4.0 & 1.0 & 4.0 & 0 \\ 0 & 0 & 4.0 & 1.0 & 3.0 \\ 0 & 0 & 0 & 4.0 & 1.0 \end{bmatrix}$$

Usual storage technique yields the following diagram:

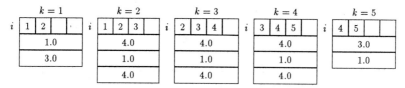

k column index, i row index

Fig. 2.3 – Usual storage technique

Since many values of the matrix A are equal a list of the different values is constructed and to each row index is assigned an index of the list of values.

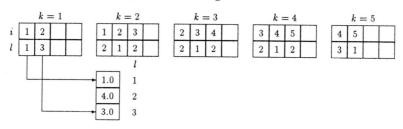

Fig. 2.4 – Modified storage technique

The saving of storage cells may be practically enormous, if, as in a linear programming example, 97000 nonzero elements assume only 4500 different values.

2.3 Unlinked lists

In general unlinked lists need even less storage cells than the techniques described above. These storage techniques are applied to avoid the use of external storages for the matrix elements, since the transportation from the external to the internal storage takes a lot of time. Additionally created nonzero elements (in various steps of the computations) are introduced by *relocating*.

Scheme 2.5

To each nonzero element correspond 2 storage cells containing the row index i and the value a_{ik}. Moreover there are items representing the current column k.

$i = 0$ denotes the end of the current column

An item $\boxed{0|0}$ denotes the end of the matrix storage.

2.3 Unlinked lists

Thus there is a total of $n + p + 1$ items, where

n : number of columns
p : number of elements $a_{ik} \neq 0$

and 1 item to indicate the end of the matrix storage.

Example 2.6
Let A be a (5×5) matrix with 7 nonzero elements

$$a_{21}, a_{41}, a_{52}, a_{13}, a_{33}, a_{24}, a_{45}.$$

Hence A is stored as the array:

$$\{0, 1|2, a_{21}|4, a_{41}|0, 2|5, a_{52}|0, 3|1, a_{13}|3, a_{33}|0, 4|2, a_{24}|0, 5|4, a_{45}|0, 0\}$$

Scheme 2.7
The information about the given matrix A is stored in 3 arrays:

V : value of elements
Z : row indices
SZ : column index pointer.

Let β be the column index of an element of A. Then $SZ(\beta) = t_\beta$ indicates that element of Z, which denotes the row index of the first nonzero element $V(A_\beta)$ of the column β. V and Z each has p elements, SZ has n elements. Thus $2p + n$ storage cells will be required.

Example 2.8
(matrix values of example (2.6))

$$\begin{aligned} V &= \{\ a_{21},\ a_{41},\ a_{52},\ a_{13},\ a_{33},\ a_{24},\ a_{45}\ \} \\ Z &= \{\ 2,\ 4,\ 5,\ 1,\ 3,\ 2,\ 4,\ \} \\ SZ &= \{\ 1,\ 3,\ 4,\ 6,\ 7\ \} \end{aligned}$$

The element a_{33} can be recovered as follows. First $SZ(3) = 4$ is determined. $Z(4) = 1$ gives the row index of the first nonzero element of the column 3. If $a_{33} \neq 0$, one of the following elements of Z equals 3; in this case $Z(5) = 3$. Hence $a_{33} = V(5)$ (i.e. $V(5)$ contains a_{33}).

Scheme 2.9
To each element $a_{ik} \neq 0$ of A is assigned a unique integer $\lambda(i, k)$, where

$$\lambda(i, k) := i + (k - 1)n. \tag{2.10}$$

Then the storage consists of 2 arrays:

$$\begin{aligned} V &= \{\text{values of } a_{ik} \neq 0\} \\ LD &= \{\text{values of } \lambda(i, k)\}, \end{aligned}$$

V and LD each having p elements. $LD(\alpha)$ is that value of $\lambda(i, k)$ corresponding to $V(\alpha) = a_{ik}$, where $\alpha = 1, \ldots, p$.

Example 2.11
(matrix values of example (2.6))

$$V = \{\ a_{21},\ a_{41},\ a_{52},\ a_{13},\ a_{33},\ a_{24},\ a_{45}\ \}$$
$$LD = \{\ 2,\ \ \ \ 4,\ \ \ \ 10,\ \ \ 11,\ \ \ 13,\ \ \ 17,\ \ \ 24,\ \}$$

The original matrix can be recovered from this storage scheme.
Equation (2.10) implies:

$$k - 1 = \frac{\lambda(i,k)}{n} - \frac{i}{n}$$
$$\Rightarrow k = \frac{\lambda(i,k)}{n} + (1 - \frac{i}{n})$$
$$\Rightarrow \frac{\lambda(i,k)}{n} \leq k < \frac{\lambda(i,k)}{n} + 1$$
$$\Rightarrow k = \min_{e \in \mathbb{Z}}\{e | e \geq \frac{\lambda(i,k)}{n}\} \qquad (2.12)$$

and

$$i = \lambda(i,k) - (k-1)n. \qquad (2.13)$$

For example (2.6):
$\lambda(i,k) = LD(5) = 13$ implies $\frac{\lambda(i,k)}{n} = \frac{13}{5}$. This yields the result $k = 3$, $i = 3$.

3 Systems of linear equations

3.1 Introductory example

As an example a problem of computer-design is chosen. The behaviour of transistor circuits is dealt with and the question arises, how the switch-mode depends on a design-parameter such that the behaviour is optimal in a sense still to be defined. This behaviour is characterized by the solution

$$w = w(t) \tag{3.1}$$

of the initial value problem

$$\dot{w} = f(t, w, p) \tag{3.2}$$

in the time interval $t_0 \leq t \leq t_F$.

A criterion-function $C = C(p) > 0$ is assigned to the equation (3.2) and the analysis in question has the purpose of determining that vector p which minimizes C:

$$\min_p(C(p)). \tag{3.3}$$

In order to optimize these problems (100 to 1000 equations of type (3.2)) very efficient numerical integration algorithms have to be provided. The following approach is proposed by [31]:

$$w_{n+1} - \alpha h \dot{w}_{n+1} = R, \tag{3.4}$$

where $R = R(w, \dot{w})$ for $t \leq t_n$, $t_{n+1} := t_n + h$.

The nonlinear system (3.4) in the implicit form of integration has to be solved by a strictly converging method, for example Newton's method:

$$(I - \alpha \cdot h \cdot J^{(k)})\Delta w = R + \alpha \cdot h \cdot \dot{w}_{n+1}^{(k)} - w_{n+1}^{(k)} \tag{3.5}$$

$$w_{n+1}^{(0)} = w_n \tag{3.6}$$

$$w_{n+1}^{(k+1)} = w_{n+1}^{(k)} + \Delta w, \tag{3.7}$$

$J = [\frac{\partial f}{\partial w}]$ is the Jacobi matrix.

In order to control the growth of the rounding errors it is important to solve first equation (3.5) for Δw and after that to determine a new w in (3.7). The form (3.5) is closely related to the method of iterative refinement of accuracy of linear algebraic equations.

The system (3.5) is of the form

$$Ax = b \tag{3.8}$$

where the matrix A is in general sparse. The system (3.8) is sparse and has to be run through repeatedly.

The matrix A is not positive definite and not diagonally dominant. The equation $Ax = b$ is solved by direct or indirect methods depending on the structure of the matrices A, i.e. the kind of sparseness.

3.2 Direct methods

Direct methods are characterized by yielding the required solution with a fixed number of arithmetical operations. This solution generally contains the usual rounding errors. If there are no rounding errors, the exact solution is obtained.

Before making use of direct methods the type of the sparse matrix should be carefully studied. In the introduction the various types were distinguished. We shall now concentrate on band matrices.

Definition 3.9
A matrix $A := [a_{ij}]$, $A \in \mathbb{R}^{n \times n}$ is called a *band matrix*, if

$$a_{ij} := 0 \quad \text{holds for} \quad i - j \geq s \quad \text{or} \quad j - i \geq t.$$

The number $m = s + t - 1$ is called the *band width*. The band is symmetric relative to the main diagonal, if $s = t$.

Example 3.10
$n = 7$, $s = 2$, $t = 3$; band width $m = 4$

$$A := \begin{bmatrix} \times & \times & \times & & & & \\ \times & \times & \times & \times & & 0 & \\ & \times & \times & \times & \times & & \\ & & \times & \times & \times & \times & \\ & & & \times & \times & \times & \times \\ & 0 & & & \times & \times & \times \\ & & & & & \times & \times \end{bmatrix}$$

3.2 Direct methods

A modification of this classical band matrix is the *circular band matrix*. It contains additional nonzero elements a_{ij} for $n + i - j < s$ and $n + j - i < t$:

$$A := \begin{bmatrix} \times & \times & \times & & & & & \times \\ \times & \times & \times & \times & & & & \\ & \times & \times & \times & \times & & & \\ & & \times & \times & \times & \times & & \\ & & & \times & \times & \times & \times & \\ \times & & & & & \times & \times & \times \\ \times & \times & & & & & \times & \times \end{bmatrix}.$$

If the band width m is significantly smaller than the order n of the matrix, the band form of the matrix is useful.

Among the direct methods for solving systems of linear equations

$$Ax = b \qquad (3.11)$$

the Gaussian elimination process and the related triangular decomposition of the matrix A plays a central role. The Gaussian elimination process appears in many algebraically equivalent variants. These variants differ in

- the way the matrix is stored

- the elimination sequence

- the precautions used to take care of rounding errors

- the method of refining solutions already determined.

Moreover, there are variants especially tailored to symmetric or positive definite matrices that need only about half of the number of storage cells.

A simple example shows how sensitively the solution in the Gaussian elimination process reacts to an interchange of rows, i.e. to the chosen pivot element. The pivot elements having values near zero are critical elements. If we use these elements, we get inexact solutions in $\mathcal{F}$ (set of fixed-point numbers) or in $\mathcal{G}$ (set of floating-point numbers).

Example 3.12
The following linear equation system is to be solved in three-figure floating-point arithmetic:

$$\begin{aligned} 0.000100 x_1 + 1.00 x_2 &= 1.00 \\ 1.00 x_1 + 1.00 x_2 &= 2.00. \end{aligned} \qquad (3.13)$$

The exact solution is

$$x_1 = 1.\overline{0001}; \quad x_2 = 0.\overline{9998},$$

hence in our arithmetic

$$x_1 = 1.00; \quad x_2 = 1.00.$$

We solve system (3.13) by the Gaussian elimination process without row interchange:
$$\begin{aligned} 0.000100 x_1 + 1.00 x_2 &= 1.00 \\ (-1.00 \cdot 10^4 + 0.00010 \cdot 10^4) x_2 &= -1.00 \cdot 10^4 + 0.0002 \cdot 10^4. \end{aligned} \tag{3.14}$$

So
$$\begin{aligned} -1.00 \cdot 10^4 x_2 &= -1.00 \cdot 10^4 \\ x_2 &= 1.00 \\ x_1 &= 0.00 \end{aligned}$$

and the pivot element is 0.000100. The solution of system (3.13) with row interchange is:
$$\begin{aligned} 1.00 x_1 + 1.00 x_2 &= 2.00 \\ (-0.000100 + 1.00) x_2 &= -0.0002 + 1. \end{aligned} \tag{3.15}$$

Hence
$$\begin{aligned} 1.00 x_2 &= 1.00 \\ x_2 &= 1.00 \\ x_1 &= 1.00. \end{aligned}$$

The pivot element being 1.00.

This example provides insight into the necessity of avoiding pivot elements of small absolute value. There are some classes of matrices (for example, positive definite matrices) for which a satisfactory analysis of rounding errors has been developed.

3.3 Analysis of the elimination process

The Gaussian elimination process for solving the system of linear equations
$$Ax = b, \tag{3.16}$$

where $b \in \mathbb{R}^n$ and $A \in \mathbb{R}^{n \times n}$, eliminates successively the unknown elements x_i until the system attains triangular form
$$Ux = c, \tag{3.17}$$

where $c \in \mathbb{R}^n$ and $U \in \mathbb{R}^{n \times n}$ is an upper triangular matrix.

The solution x of (3.17) is the same as that of the system (3.16). In case of $u_{ii} \neq 0$ for each $i = 1, 2, \ldots, n$ the solution can be easily determined:
$$x_i = (c_i - \sum_{k=i+1}^{n} u_{ik} x_k)/u_{ii} \quad \text{for } i = n, n-1, \ldots, 1.$$

The first step of this method is the elimination of the unknown x_1 in the equations $2, 3, \ldots, n$. This is in general possible only if $a_{11} \neq 0$, but there are circumstances such that if one element $a_{i1} \neq 0$, $i \in \{1, \ldots, n\}$, row interchange can produce the same effect. The matrix $A = A^{(1)}$ then passes over to a matrix

3.3 Analysis of the elimination process

$A^{(2)}$ having in its first column only zero elements below the main diagonal. Premultiplying the matrix $A^{(1)}$ by the matrix $M_1 = I - m^{(1)} e_1^T$ yields exactly the matrix $A^{(2)}$.

Matrices of the form
$$M_k = I - m^{(k)} e_k^T,$$
where
$$m^{(k)} \in \mathbb{R}^n \text{ and } e_i^T m^{(k)} = 0 \text{ for } i = 1, 2, \ldots, k$$
are called *Frobenius matrices*, or elementary lower triangular matrices.

The first step of this method produces a vector $m^{(1)}$; its components are
$$e_i^T m^{(1)} = \begin{cases} 0 & \text{for } i = 1 \\ a_{i1}/a_{11} & \text{for } i = 2, 3, \ldots, n. \end{cases}$$

Observe that this step leaves the first row of the matrix unchanged. After the first step we have with
$$b^{(2)} = M_1 b$$
the following system of equations:
$$A^{(2)} x = b^{(2)}.$$

Let us look at the kth step of the process, $k = 1, 2, \ldots, n-1$. The unknown x_k of the equations $k+1, k+2, \ldots, n$ in the system
$$A^{(k)} x = b^{(k)}$$
is to be eliminated.

Below the main diagonal of the matrix $A^{(k)}$ the columns $i = 1, 2, \ldots, k-1$ have zero elements. The elimination of the unknown x_k in the equations $k+1, k+2, \ldots, n$ is equivalent to a transformation of the matrix $A^{(k)}$ into a matrix $A^{(k+1)}$, whose column k has – contrary to $A^{(k)}$ – only zero elements below its main diagonal. This transformation of $A^{(k)}$ into $A^{(k+1)}$ is obtained by premultiplying $A^{(k)}$ by the Frobenius matrix
$$M_k = I - m^{(k)} e_k^T.$$

With $A^{(k)} = [a_{ik}^{(k)}]$ the components of the vector $m^{(k)}$ are:
$$e_i^T m^{(k)} = \begin{cases} 0 & \text{for } i = 1, 2, \ldots, k \\ a_{ik}^{(k)} / a_{kk}^{(k)} & \text{for } i = k+1, k+2, \ldots, n. \end{cases}$$

We recognize that the step k is possible only if the pivot element $a_{kk}^{(k)}$ does not vanish. As mentioned above row interchange can produce such a condition, if one $a_{ik}^{(k)} \neq 0$, for $i = k, k+1, \ldots, n$.

Let us consider the elements of the matrix
$$A^{(k+1)} = [a_{ij}^{(k+1)}],$$

where
$$A^{(k+1)} = M_k A^{(k)};$$
or, more precisely:
$$A^{(k+1)} = A^{(k)} - m^{(k)}(e_k^T A^{(k)}).$$

- The first k rows remain unchanged, since for $i = 1, 2, \ldots, k$,
$$\begin{aligned} a_{ij}^{(k+1)} &= a_{ij}^{(k)} - (e_i^T m_k) a_{kj}^{(k)} \\ &= a_{ij}^{(k)}, \end{aligned}$$
because $e_i^T m^{(k)} = 0$.

- The elements $a_{ij}^{(k+1)}$ of the columns $j = 1, 2, \ldots, k$ vanish below the main diagonal ($i > j$):
$$\begin{aligned} a_{ij}^{(k+1)} &= a_{ij}^{(k)} - (e_i^T m^{(k)}) a_{kj}^{(k)} \\ &= a_{ij}^{(k)} - \frac{a_{ik}^{(k)}}{a_{kk}^{(k)}} a_{kj}^{(k)}. \end{aligned}$$
Since $a_{ij}^{(k)} = 0$ holds for $j = 1, 2, \ldots, k-1$ and $i > j$, it follows that
$$a_{ij}^{(k+1)} = 0.$$
For $j = k$ and $i > k$ it follows that
$$a_{ik}^{(k+1)} = a_{ik}^{(k)} - \frac{a_{ik}^{(k)}}{a_{kk}^{(k)}} a_{kk}^{(k)} = 0.$$

- Thus only the elements $a_{ij}^{(k+1)}$, where $i, j > k$, are to be determined:
$$a_{ik}^{(k+1)} = a_{ij}^{(k)} - a_{ik}^{(k)} \frac{a_{kj}^{(k)}}{a_{kk}^{(k)}}.$$

So the matrix $A^{(k+1)}$ is of the form:
$$A^{(k+1)} = \begin{bmatrix} a_{11}^{(1)} & a_{12}^{(1)} & \cdots & a_{1,k}^{(1)} & a_{1,k+1}^{(1)} & \cdots & a_{1,n}^{(1)} \\ 0 & a_{22}^{(2)} & \cdots & a_{2,k}^{(2)} & a_{2,k+1}^{(2)} & \cdots & a_{2,n}^{(2)} \\ & 0 & \ddots & \vdots & \vdots & & \vdots \\ & & \ddots & a_{k,k}^{(k)} & a_{k,k+1}^{(k)} & \cdots & a_{k,n}^{(k)} \\ & & & 0 & a_{k+1,k+1}^{(k+1)} & \cdots & a_{k+1,n}^{(k+1)} \\ & & & & \vdots & \ddots & \vdots \\ 0 & 0 & \cdots & 0 & a_{n,k+1}^{(k+1)} & \cdots & a_{n,n}^{(k+1)} \end{bmatrix}.$$

3.3 Analysis of the elimination process

After $(n-1)$ steps the process terminates and we are left with the system

$$A^{(n)}x = b^{(n)},$$

where

$$A^{(n)} = M_{n-1}M_{n-2}\cdot\ldots\cdot M_1 A$$
$$b^{(n)} = M_{n-1}M_{n-2}\cdot\ldots\cdot M_1 b.$$

The matrix $A^{(n)}$ is upper triangular with $A^{(n)} = U$. The matrices M_k, $k \leq n-1$, are lower triangular matrices and so also are their products and their inverses too:

$$L_{n-1} = M_1^{-1}M_2^{-1}\cdot\ldots\cdot M_{n-1}^{-1}.$$

Theorem 3.18
If

$$L_k = M_1^{-1}M_2^{-1}\cdot\ldots\cdot M_k^{-1}, \quad \text{for } k=1,\ldots,n-1,$$

then

$$L_k = I + \sum_{i=1}^{k} m^{(i)} e_i^T.$$

Proof:
(This uses the principle of mathematical induction on k.)

$$\text{For } k=1: \quad L_1 = M_1^{-1}.$$

Since

$$\begin{aligned}(I + m^{(1)}e_1^T)M_1 &= (I + m^{(1)}e_1^T)(I - m^{(1)}e_1^T) \\ &= I - m^{(1)}(e_1^T m^{(1)})e_1^T \\ &= I\end{aligned}$$

the equation

$$L_1 = I + m^{(1)}e_1^T$$

holds.
Assuming the assertion is true for $k=1$:

$$\begin{aligned}L_{k+1} &= L_k M_{k+1}^{-1} \\ &= (I + \sum_{i=1}^{k} m^{(i)}e_i^T)(I + m^{(k+1)}e_{k+1}^T) \\ &= I + \sum_{i=1}^{k} m^{(i)}e_i^T + m^{(k+1)}e_{k+1}^T \\ &\quad + \sum_{i=1}^{k} m^{(i)}(e_i^T m^{(k+1)})e_{k+1}^T) \\ &= I + \sum_{i=1}^{k} m^{(i)}e_i^T,\end{aligned}$$

since $e_i^T m^{(k+1)} = 0$ for $i = 1, 2, \ldots, k$.

Let $a_{ik}^{(k)}/a_{kk}^{(k)} = l_{ik}^{(k)}$ for $i, k = 1, \ldots, n$. Then the matrix is of the following form:

$$L_{n-1} = \begin{bmatrix} 1 & & & & \\ l_{21}^{(1)} & 1 & & 0 & \\ l_{31}^{(1)} & l_{31}^{(2)} & \ddots & & \\ \vdots & \vdots & \ddots & 1 & \\ l_{n,1}^{(1)} & l_{n,2}^{(2)} & \cdots & l_{n,n-1}^{(n-1)} & 1 \end{bmatrix},$$

and so L_{n-1} is a lower triangular matrix. □

Define $L := L_{n-1}$. Then the given matrix A is decomposable into the product of a lower triangular matrix L and an upper triangular matrix U:

$$A = L \cdot U.$$

The original system $Ax = b$ is equivalent to the system $LUx = b$ which enables it to be decomposed into 2 staggered systems:

$$\boxed{Ax = b} \Leftrightarrow \boxed{\begin{array}{l} Ly = b \\ Ux = y \end{array}} \qquad (3.19)$$

Advantage:
If the matrix A is decomposable, we can use the known elimination process in the case of a new right-hand side $\hat{b}$ of the system (3.19) and the same matrix A; we need only apply the transformation to the new right side $\hat{b}$ (saving of operations!).

In practice such problems are often encountered where linear systems with new right-hand sides are to be solved successively.

3.4 Organization of storage

We decompose a matrix $A = [a_{ij}]$ into a lower triangular matrix

$$L = [l_{ij}], \quad \text{where } l_{ii} = 1 \text{ and } l_{ij} = 0 \text{ for } i < j$$

and an upper triangular matrix

$$U = [u_{ij}], \quad \text{where } u_{ij} = 0 \text{ for } i > j.$$

For $i = 1, 2, \ldots, n$ and $i \leq j$ we obtain:

$$a_{ij} = \sum_{k=1}^{n} l_{ik} u_{kj} = \sum_{k=1}^{i} l_{ik} u_{kj} = u_{ij} + \sum_{k=1}^{i-1} l_{ik} u_{kj}.$$

3.4 Organization of storage

Hence,

$$u_{ij} = a_{ij} - \sum_{k=1}^{i-1} l_{ik} u_{kj} \text{ for } j = i, i+1, \ldots, n$$

and by analogy to the above we get

$$l_{ji} = (a_{ji} - \sum_{k=1}^{i-1} l_{jk} u_{ki})/u_{ii} \text{ for } j = i+1, \ldots, n.$$

These equations show that each element a_{ij} of the matrix A is needed exactly once, namely for the determination of either the corresponding element u_{ij}, $i \leq j$, of the matrix U or the corresponding element l_{ij}, $i > j$, of the matrix L (the elements $l_{ii} = 1$ are known).

Consequently, if an element a_{ij} has been used once, the decomposition process will never require this element again; thus its matrix position is 'free' and we can replace the element with u_{ij} or l_{ij}, respectively. We do not need any further storage cell and at the end of the process we obtain a matrix $A = [a_{ij}]$, where

$$a_{ij} = \begin{cases} u_{ij} & \text{for } i \leq j \\ l_{ij} & \text{for } i > j \end{cases}.$$

Finally let us consider the Theorem asserting the uniqueness of the triangular decomposition of a matrix.

Theorem 3.20
Let $A \in \mathbb{R}^{n \times n}$ and assume $A_k \in \mathbb{R}^{k \times k}$ to be a submatrix of A that consists of the first k rows and first k columns of A. Furthermore, let $det(A_k) \neq 0$ for $k = 1, 2, \ldots, n-1$. Then there exists an unique lower triangular matrix $L = [l_{ij}]$ having only unities in its diagonal and a unique upper triangular matrix $U = [u_{ij}]$ with $A = L \cdot U$.

Proof:
(by principle of mathematical induction)

$$\text{For } n = 1: \quad A = [a_{11}] = 1 \cdot [u_{11}].$$

Assuming the assertion is true for $n = k - 1$, it follows for $n = k$:

$$A_k = \begin{bmatrix} A_{k-1} & b \\ c^T & a_{kk} \end{bmatrix} \; ; \; L_k = \begin{bmatrix} L_{k-1} & 0 \\ m^T & 1 \end{bmatrix} \; ; \; U_k = \begin{bmatrix} U_{k-1} & r \\ 0 & u_{kk} \end{bmatrix}$$

where

$$b, c, m, r \in \mathbb{R}^{k-1}$$

$$a_{kk}, u_{kk} \in \mathbb{R}$$

$$A_{k-1}, L_{k-1}, U_{k-1} \in \mathbb{R}^{(k-1) \times (k-1)}.$$

Then $A_k = L_k \cdot U_k$ implies

$$\begin{aligned} A_{k-1} &= L_{k-1} \cdot U_{k-1} \\ b &= L_{k-1} \cdot r \\ c^T &= m^T \cdot U_{k-1} \\ a_{kk} &= m^T r \cdot u_{kk}. \end{aligned}$$

The Theorem 3.20 is proved, if r, m, u_{kk} can be determined uniquely. Since

$$det(L_{k-1}) = 1$$

holds, the equation $L_{k-1}r = b$ yields a unique solution r. Since

$$det(A_{k-1}) = det(L_{k-1}) \cdot det(U_{k-1}) = det(U_{k-1}) \neq 0$$

holds, the equation $m^T U_{k-1} = c^T$ leads to the unique solution m. Hence it follows that $u_{kk} = a_{kk} - m^T r$ is uniquely determinable. □

Thus the condition $det(A_k) \neq 0$, for $k = 1, 2, \ldots, n$ is equivalent to the existence of the triangular factorization of the matrix A without pivoting.

3.5 Minimizing the fill-in

Usually the classical Gaussian elimination process generates a new nonzero element in place of a zero element.

For example, if in the kth step the elements $a_{kk}^{(k)}, a_{kj}^{(k)}, a_{ik}^{(k)}$ are all nonzero and $a_{ij}^{(k)} = 0$, where $i, j > k$, then

$$a_{ij}^{(k+1)} := -\frac{a_{ik}^{(k)}}{a_{kk}^{(k)}} \cdot a_{kj}^{(k)} \neq 0, \tag{3.21}$$

follows, i.e. under the above-mentioned conditions a zero element in (i, j)th position of $A^{(k)}$ becomes a nonzero element in $A^{(k+1)}$.

Definition 3.22
The total number of all such elements changing from zero in $A^{(k)}$ to nonzero in $A^{(k+1)}$ is called the *local fill-in*.

Instead of choosing $a_{kk}^{(k)}$ as the pivot element in the kth step, suppose that another nonzero element $a_{st}^{(k)}$, $s \geq k$, $t \geq k$, is selected as pivot element; then the kth and the sth row, as well as the kth and tth column of $A^{(k)}$ have to be interchanged before calculating $A^{(k+1)}$. So it is necessary to evaluate

$$A^{(k+1)} := M_k \hat{A}^{(k)}, \quad k = 1, 2, \ldots, n, \tag{3.23}$$

where

$$\hat{A}^{(k)} := P_k A^{(k)} Q_k. \tag{3.24}$$

Both P_k and Q_k are permutation matrices interchanging the kth, sth row and the kth, tth column, respectively. M_k is an elementary lower triangular matrix:

3.5 Minimizing the fill-in

$$M_k = \begin{bmatrix} 1 & & & & & & \\ & \ddots & & & 0 & & \\ & & \ddots & & & & \\ & & & 1 & & & \\ & & & \times & \ddots & & \\ & & & \vdots & & \ddots & \\ & & & \times & & & 1 \end{bmatrix} \leftarrow k\text{th row}$$

$$\uparrow$$
$$k\text{th column}$$

M_k is given by

$$M_k := I_n - m^{(k)} e_k^T, \quad k = 1, 2, \ldots, n; \tag{3.25}$$

and the elements of the column vector $m^{(k)}$ are defined as follows:

$$m_i^{(k)} := \begin{cases} 0 & \text{for } i \leq k \\ \hat{a}_{ik}^{(k)}/\hat{a}_{kk}^{(k)} & \text{for } i > k. \end{cases} \tag{3.26}$$

Let $\epsilon > 0$ be the pivot tolerance, i.e. the set

$$\mathcal{T} := \{|\, a_{st}^{(k)} \,| > \epsilon \text{ with } s \geq k, t \geq k\} \tag{3.27}$$

is considered. That pivot element is selected from $\mathcal{T}$ which minimizes the local fill-in.

Definition 3.28
Let B_k be the matrix that is obtained, if the nonzero elements in the last $n-k+1$ rows and columns of $A^{(k)}$ are replaced by unity.

Theorem 3.29
Choose $a_{i+k-1,j+k-1}$ as pivot element in the kth step of the Gaussian elimination process. Then the local fill-in is given by the (i,j)th element of the matrix G_k, where

$$G_k := B_K \bar{B}_k^T B_k \tag{3.30}$$

and $\bar{B}_k^T$ is the transpose of the matrix obtained when each zero element of B_k is replaced by unity and vice versa.

Proof:
At the kth step of the Gaussian elimination process let us choose as pivot element the element $a_{i+k-1,j+k-1}^{(k)}$. The equations

$$a_{k+l-1,k+m-1}^{(k+1)} = a_{k+l-1,k+m-1}^{(k)} - \frac{a_{k+i-1,k+m-1}^{(k)} \cdot a_{k+l-1,k+j-1}^{(k)}}{a_{k+i-1,k+j-1}^{(k)}}$$

where $l, m = 1, 2, \ldots, n - k + 1$, $l \neq i$, indicate the transformations. A fill-in at the position $(k + l - 1, k + m - 1)$ is effected if and only if

$$a^{(k)}_{k+l-1, k+m-1} = 0$$

$$a^{(k)}_{k+i-1, k+m-1} \neq 0$$

$$a^{(k)}_{k+l-1, k+j-1} \neq 0.$$

That is if $b_{lm} = 0$ and $b_{im} = b_{lj} = 1$, or $b_{lj}(1 - b_{lm})b_{im} = 1$. Thus the total fill-in is

$$\sum_{l=1}^{n-1+k} \sum_{m=1}^{n-1+k} b_{lj}(1 - b_{lm})b_{im}.$$

But this is exactly the element g_{ij} of the matrix $G_k = B_k \bar{B}_k^T B_k$, as was to be proved. □

The Theorem 3.29 implies the following Corollary (see [57]):

Corollary 3.31
If $a^{(k)}_{st}$ is chosen as the pivot element in the kth step of the Gaussian elimination process, where $s = \alpha + k - 1$, $t = \beta + k - 1$ and α, β are so chosen that

$$g^{(k)}_{\alpha, \beta} = \min_{i,j} e_i^T G_k e_j$$

for all

$$| a^{(k)}_{i+k-1, j+k-1} | > \epsilon, \quad (\epsilon : \text{pivot tolerance}),$$

then the local fill-in will be minimized.

3.6 Stability considerations

In 1974 a method for the restriction of numerical instability was suggested in [9]. The starting point is the task of solving systems of linear equations

$$Ax = b, \tag{3.32}$$

where A is a sparse matrix.

In applied mathematics a substantial number of systems like (3.32) have to be solved where the matrices A have the same sparseness pattern. In Gaussian elimination it is important to implement row and column interchange in such a way as to preserve the sparseness properly. Unfortunately this is often far more expensive than performing a fixed pattern of elimination. It is therefore desirable to implement the same interchanges as often as possible.

To simplify notation, we define A to be the matrix after the interchanges of rows and columns. We look at the case of numerical instability arising from the resulting computation.

3.6 Stability considerations

Wilkinson's error theory and error analysis [58] shows that, without imposing any restricting condition on the sizes of the elements of L, the calculated triangular factors L and U of A satisfy the equation

$$A + F = L \cdot U, \qquad (3.33)$$

where F is the matrix of errors whose elements f_{ij} satisfy the inequality

$$|f_{ij}| \leq 3.01 \cdot \epsilon \, \rho \, m_{ij} \qquad (3.34)$$

with

ϵ : relative accuracy of the arithmetic in use

ρ : $\max_{i,j,k} |a_{ij}^{(k)}|$ (largest element in any of the intermediate matrices encountered in the elimination process)

m_{ij} : numbers of multiplications required in the calculation of l_{ij}, $(i > j)$ or u_{ij}, $(i \leq j)$ (see [58] under rounding errors).

If rows and columns are interchanged the size of ρ is normally kept under control.

Neglecting rounding errors, the elements of the kth reduced matrix in the Gaussian elimination process are given by

$$a_{ij}^{(k)} := a_{ij} - \sum_{m=1}^{k} l_{im} u_{mj}, \quad k < i \leq n, \; k < j \leq n. \qquad (3.35)$$

Using Hölder's inequality with $\frac{1}{p} + \frac{1}{q} = 1$ the following estimate can be made:

$$|a_{ij}^{(k)}| \leq |a_{ij}| + \|(l_{i1}, l_{i2}, \ldots, l_{ik})\|_p \cdot \|(u_{1j}, u_{2j}, \ldots, u_{kj})\|_q. \qquad (3.36)$$

This inequality can be weakened [5]:

$$|a_{ij}^{(k)}| \leq \max_{i,j} |a_{ij}| + \max_{i} \|(l_{i1}, \ldots, l_{i,i-1})\|_p \cdot \max_{j} \|(u_{1j}, \ldots, u_{j-1,j})\|_q. \qquad (3.37)$$

Due to the advantageous form of L and U the estimation (3.37) can be calculated economically in each of the cases $p = 1, 2, \infty$; to compute the norm, only one reference for each nonzero off-diagonal element of L and U is needed.

In case of symmetric matrices (real or complex) the well-known decomposition of these matrices gives

$$a_{ij}^{(k)} := a_{ij} - \sum_{m=1}^{k} l_{im} d_m l_{mj}, \quad k < i \leq n, \; k < j \leq n. \qquad (3.38)$$

The computational effort and the storage is halved. It was shown by [58] that in case of real and positive definite matrices

$$\mid a_{ij}^{(k)} \mid \leq \max_{i,j} \mid a_{ij} \mid$$

and that the corresponding symmetric algorithm is applicable only in this case.

According to experience monitoring the stability is possible also in more general cases.

For $p = 2$ the inequality

$$\mid a_{ij}^{(k)} \mid \leq \max_{i,j} \mid a_{ij} \mid + \max_{i} \sum_{m=1}^{i-1} \mid l_{im}^2 d_m \mid \qquad (3.39)$$

is obtained. The same bounds hold also for Hermitian matrices.

In order to improve the exactness of the solution we make use of the well-tried iteration

$$\begin{aligned} LU\delta^{(k)} &= b - Ax^{(k)} \\ x^{(k+1)} &= x^{(k)} - \delta^{(k)} \end{aligned} \qquad k = 0, 1, 2, \ldots \qquad (3.40)$$

commencing with $x^{(0)} = 0$. The size of changes $\delta^{(k)}$ gives a good guide to the level of uncertainty in the approximate solution $x^{(k)}$.

The bounds (3.36) and (3.37) can be expected to be quite realistic if the values of the elements of L and U do not oscillate too much. Otherwise the bounds will be poor.

4 Computation of the inverse

In 1975 a recursive algorithm for computing the inverse of a sparse matrix A from LU factors was developed in [10]. A particular application of this algorithm is the computation of certain elements of the inverse of a sparse matrix A.

The following algorithm does not involve the determination of L^{-1} and thus it makes more use of the sparseness. There is a certain dependency relationship among the elements of the inverse permitting an efficient determination of a subset of these inverse elements.

Let A be a nonsingular $(n \times n)$ matrix. We assume the elements of the matrix A to be appropriately ordered for factorization, and let

$$A = L \cdot D \cdot U. \tag{4.1}$$

L is a lower triangular matrix all of whose diagonal elements are unity, and U is an upper triangular matrix having also unit elements in its diagonal. D is a diagonal matrix:

$$L := \begin{bmatrix} 1 & & 0 \\ & \ddots & \\ l_{ij} & & 1 \end{bmatrix}, \quad U := \begin{bmatrix} 1 & & u_{ij} \\ & \ddots & \\ 0 & & 1 \end{bmatrix}, \quad D := \begin{bmatrix} d_1 & & 0 \\ & \ddots & \\ 0 & & d_n \end{bmatrix}.$$

Let
$$Z := A^{-1}. \qquad (4.2)$$

Then, according to [56] we have the formulae:
$$Z := D^{-1}L^{-1} + (I - U)Z \qquad (4.3)$$

$$Z := U^{-1}D^{-1} + Z(I - L). \qquad (4.4)$$

$(I-U)$ and $(I-L)$ are strictly upper and lower triangular matrices respectively.

By using (4.3) for the elements of the upper triangular matrix of Z and (4.4) for the elements of the lower triangular matrix of Z, L^{-1} and U^{-1} do not enter into the computation of the elements z_{ij}. Thus these formulae provide a means of computing the elements z_{ij} in terms of previously determined elements of Z.

In the special case of a symmetric matrix A (i.e. $A = A^T$), this matrix is factorized into
$$A = L \cdot D \cdot L^T \qquad (4.5)$$

and so we obtain an equation for Z:
$$Z = D^{-1}L^{-1} + (I - L^T)Z. \qquad (4.6)$$

Definition 4.7
In order to derive the dependency relationship existing among the elements of the inverse the *adjacency matrix* C, associated with $A = L \cdot U$, is defined by

$$c_{ij} = \begin{cases} 1 & \text{for } l_{ij} \neq 0 \text{ or } u_{ij} \neq 0 \\ 0 & \text{otherwise,} \end{cases} \qquad (4.8)$$

i.e. C has the sparseness pattern of $L \cdot U$.

When working with sparse matrices it has to be assumed necessarily that any l_{ij} or u_{ij} to be computed is treated as nonzero, even if it becomes zero due to numerical cancellation, i.e. it is assumed:

$$c_{ki} = c_{ij} = 1 \Rightarrow c_{kj} = 1 \text{ for } k,j > i. \qquad (4.9)$$

Note: $c_{ii} = 1$ for each i, $i = 1, \ldots, n$.

Equation (4.3) implies

$$z_{ij} := -\sum_{k=i+1}^{n} u_{ik} z_{kj} \text{ for } i \neq j \qquad (4.10)$$

and

$$z_{ii} := \frac{1}{d_{ii}} - \sum_{k=i+1}^{n} u_{ik} z_{kj} \text{ for } i = j. \qquad (4.11)$$

For the computation of z_{ij} start in the (n,n) position and compute the elements working upward to z_{ij} according to (4.10) and (4.10).

4 Computation of the inverse

Example 4.12

$$A := \begin{bmatrix} 1 & 0 & 0 & 2 \\ 0 & 2 & 2 & 0 \\ 1 & 0 & 1 & 2 \\ 1 & 0 & 0 & 4 \end{bmatrix}$$

If A is decomposed into $L\,D\,U$ we have:

$$L := \begin{bmatrix} 1 & 0 & 0 & 0 \\ 0 & 1 & 0 & 0 \\ 1 & 0 & 1 & 0 \\ 1 & 0 & 0 & 1 \end{bmatrix} \quad D := \begin{bmatrix} 1 & 0 & 0 & 0 \\ 0 & 2 & 0 & 0 \\ 0 & 0 & 1 & 0 \\ 0 & 0 & 0 & 2 \end{bmatrix} \quad U := \begin{bmatrix} 1 & 0 & 0 & 2 \\ 0 & 1 & 1 & 0 \\ 0 & 0 & 1 & 0 \\ 0 & 0 & 0 & 1 \end{bmatrix}.$$

Computation of z_{23} and z_{11}:

$$\begin{aligned} z_{23} &= -\sum_{k=4}^{4} u_{2k} z_{k3} = -(u_{23} z_{33} + u_{24} z_{43}) = -z_{33} \\ z_{33} &= (1/d_{33}) - u_{34} z_{43} = 1 \end{aligned}$$

implying that $z_{23} = -1$.

$$\begin{aligned} z_{11} &= (1/d_{11}) - \sum_{k=2}^{4} u_{1k} z_{k1} = 1 - (u_{12} z_{21} + u_{13} z_{31} + u_{14} z_{41}) \\ &= 1 - 2 z_{41} \\ z_{41} &= -\sum_{k=2}^{4} z_{4k} l_{k1} = -(z_{42} l_{21} + z_{43} l_{31} + z_{44} l_{41}) \\ &= -z_{43} - z_{44} \\ z_{44} &= 1/d_{44} = 1/2 \\ z_{43} &= -z_{44} l_{43} = 0 \end{aligned}$$

Thus:

$$\begin{aligned} z_{41} &= -\tfrac{1}{2} \\ z_{11} &= 1 - (-1) = 2. \end{aligned}$$

In sensitivity computation for the problem $Ax = b$ the sensitivity of x_i to a_{jk} is given by

$$\frac{\partial x_i}{\partial a_{jk}} = x_k z_{ik}.$$

A further example is the condition number of a sparse symmetric positive definite matrix which may be approximated by using the diagonal elements of the inverse matrix.

5 Block matrices

5.1 Block elimination of linear sparse systems

Unless otherwise stated, $A := [a_{ik}]$, $A \in \mathbb{R}^{n \times n}$, is assumed to be a nonsymmetric sparse matrix.

Assumption: The matrix A possesses a triangular decomposition. From the practical point of view the matrix should allow an acceptable factorization if Gaussian elimination without interchange is applied to A.

This is the case when A is

1. weakly diagonally dominant (see also Definition 6.7), i.e.

$$| a_{ii} | \geq \sum_{\substack{j=1 \\ j \neq i}}^{n} | a_{ij} | \quad \text{for all } i = 1, 2, \ldots, n;$$

or

2. symmetric and positive definite, i.e. all eigenvalues λ_i are positive.

In these two cases Gaussian elimination with the natural pivot elements can be carried out without loss of accuracy.

Note 5.1
Such systems often arise when solving special elliptic boundary value problems by the methods of finite differences and finite elements.

5.1 Block elimination of linear sparse systems

Definition 5.2
A matrix A is said to be *reducible* if there exists a permutation matrix P such that
$$PAP^T = \begin{bmatrix} A_{11} & A_{12} \\ 0 & A_{22} \end{bmatrix};$$
otherwise A is said to be *irreducible* (see also Definition (6.10)).

To start with, the problem
$$Ax = b \tag{5.3}$$
with $n = p + q$, $p, q > 0$, is considered and decomposed into
$$\begin{bmatrix} D & G \\ B^T & F \end{bmatrix} \cdot \begin{bmatrix} x_1 \\ x_2 \end{bmatrix} = \begin{bmatrix} b_1 \\ b_2 \end{bmatrix}, \tag{5.4}$$
where $x_1, b_1 \in \mathbb{R}^p$, $x_2, b_2 \in \mathbb{R}^q$. D is a $(p \times p)$ matrix, F a $(q \times q)$ matrix and both are nonsingular. B and G are $(p \times q)$ matrices.

Furthermore, the matrices B and G are assumed to be non-null, for otherwise (5.4) can be reduced to a smaller system. This assumption is fulfilled if A is irreducible.

Theorem 5.5
If A in (5.3) with the property (5.4) is decomposable into $A = L_a U_a$ by Gaussian elimination, where
$$L_a = \begin{bmatrix} 1 & & 0 \\ & \ddots & \\ l_{ij}^{(a)} & & 1 \end{bmatrix}, \quad U_a = \begin{bmatrix} u_{11}^{(a)} & & u_{ij}^{(a)} \\ & \ddots & \\ 0 & & u_{nn}^{(a)} \end{bmatrix},$$
then
$$A = L_a U_a = \begin{bmatrix} L & 0 \\ W^T & L_e \end{bmatrix} \cdot \begin{bmatrix} U & V \\ 0 & U_e \end{bmatrix}, \tag{5.6}$$
where
$$D := LU \tag{5.7}$$
$$V := L^{-1} G \tag{5.8}$$
$$W^T := B^T U^{-1} \tag{5.9}$$
$$L_e U_e := \tilde{F} := F - W^T V. \tag{5.10}$$

Proof:
By multiplication (5.6) yields
$$D = LU$$
$$G = LV$$
$$B^T = W^T U$$
$$F = W^T V + L_e U_e$$
and hence immediately (5.7) to (5.10). □

Then with (5.7) to (5.10) the factorization (5.6) can be carried out block by block. However, according to [4], this leads to no significant improvement compared with Gaussian elimination.

Concerning the above-mentioned problem, the following block factorization can also be considered:

$$A = \begin{bmatrix} D & 0 \\ B^T & L_e \end{bmatrix} \cdot \begin{bmatrix} I & \tilde{V} \\ 0 & U_e \end{bmatrix}, \qquad (5.11)$$

where

$$\tilde{V} = U^{-1}V, \qquad (5.12)$$

or the factorization

$$A = \begin{bmatrix} I & 0 \\ \tilde{W}^T & L_e \end{bmatrix} \cdot \begin{bmatrix} D & G \\ 0 & U_e \end{bmatrix}, \qquad (5.13)$$

where

$$\tilde{W}^T = W^T L^{-1}. \qquad (5.14)$$

Introduction of abbreviations:

Factorization (5.6) is denoted by F_1.
Factorization (5.11) is denoted by F_2.
Factorization (5.13) is denoted by F_3.

All factorizations are assumed to allow the usual decomposition (by Gaussian elimination) of D and F, and only the (2×2) block decomposition is considered. More general block decompositions are possible.

The three different block decompositions F_1, F_2, F_3 have been investigated in [17] to see whether significant reductions in computational and storage costs might be achieved. In many cases F_2 and F_3 led to a significant improvement compared with F_1.

Note 5.15
In some cases the equations may be ordered so that only a few blocks are generated. The block system can be ordered in such a way as to give little or no block fill-in.

The factorizations F_1, F_2, F_3 yield the following *backward substitution* steps:

F_1	F_2	F_3	
$Ly_1 = b_1$	$LUy_1 = b_1$	$y_1 = b_1$	(5.16a)
$\bar{b}_2 = b_2 - W^T y_1$	$\bar{b}_2 = b_2 - B^T y_1$	$\bar{b}_2 = b_2 - \tilde{W}^T y_1$	(5.16b)
$L_e y_2 = \bar{b}_2$	$L_e y_2 = \bar{b}_2$	$L_e y_2 = \bar{b}_2$	(5.16c)
$U_e x_2 = y_2$	$U_e x_2 = y_2$	$U_e x_2 = y_2$	(5.16d)
$\bar{y}_1 = y_1 - V x_2$	$\bar{y}_1 = y_1 - \tilde{V} x_2$	$\bar{y}_1 = y_1 - G x_2$	(5.16e)
$U x_1 = \bar{y}_1$	$x_1 = \bar{y}_1$	$LU x_1 = \bar{y}_1$	(5.16f)

(5.16)

(5.16d) and (5.16f) give the solution $x = \begin{bmatrix} x_1 \\ x_2 \end{bmatrix}$.

Note 5.17
Bunch and Rose [4] have made the important observation that if B, G, L, U are sufficiently sparse in comparison with W and V, the number of operations of back substitution in the factorization F_1 can be reduced, W and V being given by their definitions above.

Example 5.18
The factorization F_1 is to be applied to the system $Ax = b$ (see Example 6.47):

$$A := \begin{bmatrix} 4 & -1 & -1 & 0 \\ -1 & 4 & 0 & -1 \\ -1 & 0 & 4 & -1 \\ 0 & -1 & -1 & 4 \end{bmatrix}, \quad b := \begin{bmatrix} 0 \\ 0 \\ 1000 \\ 1000 \end{bmatrix}.$$

The matrix A is decomposed into:

$$D = \begin{bmatrix} 4 & -1 \\ -1 & 4 \end{bmatrix}, \quad G = \begin{bmatrix} -1 & 0 \\ 0 & -1 \end{bmatrix}, \quad B^T = \begin{bmatrix} -1 & 0 \\ 0 & -1 \end{bmatrix}, \quad F = \begin{bmatrix} 4 & -1 \\ -1 & 4 \end{bmatrix}.$$

The factorization yields:

$$L = \begin{bmatrix} 1 & 0 \\ -\frac{1}{4} & 1 \end{bmatrix}, \quad U = \begin{bmatrix} 4 & -1 \\ 0 & 3\frac{3}{4} \end{bmatrix}, \quad W^T = \begin{bmatrix} -\frac{1}{4} & -\frac{1}{15} \\ 0 & -\frac{4}{15} \end{bmatrix},$$

$$L_e = \begin{bmatrix} 1 & 0 \\ -\frac{2}{7} & 1 \end{bmatrix}, \quad U_e = \begin{bmatrix} 3\frac{11}{15} & -1\frac{1}{15} \\ 0 & 3\frac{3}{7} \end{bmatrix}, \quad V = \begin{bmatrix} -1 & 0 \\ -\frac{1}{4} & -1 \end{bmatrix}.$$

By backward substitution according to (5.16a) to (5.16f) the solution

$$x = \begin{bmatrix} 125 \\ 125 \\ 375 \\ 375 \end{bmatrix}$$

is obtained.

5.2 Property $\mathcal{P}$

In the preceding discussions the sparseness of $L, U, \tilde{W}, V$ and $\tilde{V}$ entered into the calculations without taking into account the structure of the matrices.

We shall now introduce a sparseness condition called the *propagation property* or, briefly, *property* $\mathcal{P}$. In these definitions and theorems the exclusion of numerical cancellation is a global precondition. Later it will be shown (see Note 5.36) that this precondition may create problems.

If D possesses property $\mathcal{P}$, then the structures of $W, \tilde{W}, V$ and $\tilde{V}$ are completely determined by the structures of B and G.

Definition 5.19
A nonsingular lower or upper $(n \times n)$ triangular matrix L or U possesses *property* $\mathcal{P}$ if $l_{i,i-1} \neq 0$, $2 \leq i \leq n$, or $u_{i,i+1} \neq 0$, $1 \leq i \leq (n-1)$, respectively.

Definition 5.20
A nonsingular $(n \times n)$ matrix M with the decomposition $M = LU$ is called M *with property* $\mathcal{P}$, if L and U possess property $\mathcal{P}$.

If M possesses property $\mathcal{P}$ it does not follow that PMQ also in all cases possesses property $\mathcal{P}$, where P, Q are $(n \times n)$ permutation matrices.

Example 5.21

$$M = \begin{bmatrix} \times & \times & 0 \\ \times & \times & \times \\ 0 & \times & \times \end{bmatrix}, \quad LU = \begin{bmatrix} \times & 0 & 0 \\ \times & \times & 0 \\ 0 & \times & \times \end{bmatrix} \cdot \begin{bmatrix} \times & \times & 0 \\ 0 & \times & \times \\ 0 & 0 & \times \end{bmatrix}.$$

$M = LU$ possesses property $\mathcal{P}$. If the final rows and columns of M are interchanged, the corresponding LU decomposition no longer possesses property $\mathcal{P}$.

Note 5.22
$f(x) = r$ denotes the index of the first nonzero component of x. $\bar{f}(x) = t$ denotes the index of the final nonzero component of x.

Theorem 5.23
Let L be a nonsingular lower $(n \times n)$ triangular matrix possessing property $\mathcal{P}$. Let x be the solution of the system $Lx = b$, where $b_i = 0$ for $1 \leq i < k$ and $b_k \neq 0$, i.e. $k = f(b)$. Then

$$x_i = 0 \quad \text{for} \quad 1 \leq i < k \quad \text{and} \quad x_i \neq 0, \ k \leq i \leq n.$$

Proof:
(by the principle of mathematical induction).
Since L is nonsingular, i.e. $l_{ii} \neq 0$ for $1 \leq i \leq n$, the following holds:

$$b_i = 0, \ 1 \leq i \leq k, \quad \text{implies} \quad x_i = 0, \ 1 \leq i \leq k.$$

Furthermore,

$$x_k = \frac{b_k}{l_{kk}} \neq 0, \quad \text{since} \ b_k \neq 0.$$

It remains to show that:

$$x_i \neq 0 \ \Rightarrow \ x_{i+1} \neq 0, \ i \geq k.$$

We have

$$x_{i+1} = (b_{i+1} - l_{i+1,i} x_i - \sum_{j=1}^{i-1} l_{i+1,j} x_j) / l_{i+1,i+1}.$$

Since numerical cancellation, as assumed at the beginning, is excluded, $l_{i+1,i} \neq 0$,

$x_i \neq 0$ implies $x_{i+1} \neq 0$. $\square$

Note 5.24
Let $n(\Delta)$ denote the number of nonzero components of Δ, where Δ is a matrix or vector.

Corollary 5.25
Let L be a nonsingular lower $(n \times n)$ triangular matrix possessing property $\mathcal{P}$. Let m_{ij} be the (i,j)th component of M, where M is a $(n \times r)$ matrix satisfying the r triangular systems $LM = \bar{M}$. Then

$$m_{ij} = 0, \quad 1 \leq i < f(\bar{m}_{.j}), \quad j = 1,\ldots,r \tag{5.26}$$

$$m_{ij} \neq 0, \quad f(\bar{m}_{.j}) \leq i \leq n, \quad j = 1,\ldots,r \tag{5.27}$$

$$n(M) = r(n+1) - \sum_{j=1}^{r} f(\bar{m}_{.j}). \tag{5.28}$$

Proof:
(5.26) and (5.27) are obtained by applying Theorem 5.23 to the columns. (5.27) implies (5.28):

$$n(M) = \sum_{j=1}^{r}(n - f(\bar{m}_{.j}) + 1)$$

$$= r(n+1) - \sum_{j=1}^{r} f(\bar{m}_{.j})$$

□

Theorem 5.29
Let U be an upper nonsingular $(n \times n)$ triangular matrix possessing property $\mathcal{P}$. Let x be the solution of $Ux = b$, $b_i = 0$ for $k < i \leq n$, $b_k \neq 0$, i.e. $k = \bar{f}(b)$. Then, if numerical cancellation is excluded,

$$x_i \neq 0 \quad \text{for} \quad 1 \leq i \leq k$$

and

$$x_i = 0 \quad \text{for} \quad k < i \leq n.$$

Proof:
Analogous to Theorem 5.23. □

Corollary 5.30
Let U be an upper $(n \times n)$ triangular matrix possessing property $\mathcal{P}$. Let m_{ij} be the (i,j)th component of M, where M is a $(n \times r)$ matrix satisfying the r triangular systems $UM = \bar{M}$. Then

$$m_{ij} \neq 0, \quad 1 \leq i \leq \bar{f}(\bar{m}_{.j}), \quad j = 1,\ldots,r$$

$$m_{ij} = 0, \quad \bar{f}(\bar{m}_{.j}) < i \leq n, \quad j = 1,\ldots,r$$

$$n(M) = \sum_{i=1}^{r} \bar{f}(\bar{m}_{.j}).$$

Proof:
The proof follows directly from Theorem 5.29 and is analogous to the proof of Corollary 5.25. □

Theorem 5.31
Let M be a $(n \times n)$ matrix possessing property $\mathcal{P}$ and $M = LU$. Let x be the solution of $(LU)x = b$, where $b_i \neq 0$ for at least one i. Then $n(x) = n$, i.e. x is a *dense vector*. [2]

Proof:
Determine x from
$$Ly = b$$
$$Ux = y.$$

According to Theorem 5.23 the final component of y is not zero; then Theorem 5.29 states that x is dense. □

Corollary 5.32
The inverse of the lower (upper) $(n \times n)$ triangular matrix possessing property $\mathcal{P}$ is a *dense lower (upper) triangular matrix*.

Corollary 5.33
The inverse of an $(n \times n)$ matrix possessing property $\mathcal{P}$ is dense.

Again the two Corollaries hold only under the condition that numerical cancellation is excluded. The proofs are direct applications of Corollaries 5.25 and 5.30.

Note 5.34
In the three decompositions F_1, F_2, F_3, the matrix D is first decomposed into LU. Then the following applies to F_1: If L possesses property $\mathcal{P}$, then because of Corollary 5.25 and the structure of G, the structure of V is known, since, by (5.8),
$$V = L^{-1}G \Leftrightarrow LV = G.$$

If U possesses property $\mathcal{P}$, then by Corollary 5.25 and the structure of B, the structure of W is known, since, by (5.9),
$$W^T = B^T U^{-1} \Leftrightarrow U^T W = B,$$

where U^T is a lower triangular matrix. Similar considerations hold for F_2 and F_3.

Note 5.35
When a matrix M is irreducible and numerical cancellation is excluded, then the inverse M^{-1}, calculated by Gaussian elimination, is a dense matrix.

---
[2] What are here described as *dense* vectors and matrices are also referred to in the literature as *full*. This indicates the antithesis of *sparse*, namely most, if not all, elements are nonzero.

Note 5.36
The following example is given to illustrate the problematic nature of the assumption that numerical cancellation is excluded.

Example 5.37

$$\begin{bmatrix} 1 & 0 & 0 \\ 1 & 1 & 0 \\ 1 & 1 & 1 \end{bmatrix} \cdot \begin{bmatrix} 0 \\ 1 \\ 0 \end{bmatrix} = \begin{bmatrix} 0 \\ 1 \\ 1 \end{bmatrix}$$

The matrix possesses property $\mathcal{P}$ und $b_2 \neq 0$. According to Theorem 5.23 $x_2 \neq 0$, $x_3 \neq 0$ should hold. As stated in Corollary 5.32 the inverse of this matrix, a lower triangular matrix, is not a dense matrix:

$$\begin{bmatrix} 1 & 0 & 0 \\ 1 & 1 & 0 \\ 1 & 1 & 1 \end{bmatrix}^{-1} = \begin{bmatrix} 1 & 0 & 0 \\ -1 & 1 & 0 \\ 0 & -1 & 1 \end{bmatrix}.$$

5.3 Direct algorithms with block matrices

The development of direct algorithms for solving sparse matrix problems is still in its infancy since the problem of the fill-in is not yet solved; one constantly aims at preserving the sparseness.

The classical procedure for solving

$$Ax = b; \quad A := [a_{ik}], \quad A \in \mathbb{R}^{n \times n}, \quad x, b \in \mathbb{R}^n, \tag{5.38}$$

falls into three main steps:

- triangularization of A
- backward substitution
- iterative refinement of accuracy.

The decomposition of A yields

$$A := PLUQ, \tag{5.39}$$

where P and Q are permutation matrices, L is a lower and U is an upper triangular matrix of the following form:

$$L := \begin{bmatrix} 1 & & 0 \\ & \ddots & \\ l_{ik} & & 1 \end{bmatrix}, \quad U := \begin{bmatrix} u_{11} & & u_{ik} \\ & \ddots & \\ 0 & & u_{nn} \end{bmatrix}.$$

The solution of (5.38) is obtained by the following easily solvable subsystems:

$$\begin{aligned} Px^{(1)} &= b \\ Lx^{(2)} &= x^{(1)} \\ Ux^{(3)} &= x^{(2)} \\ Qx &= x^{(3)}. \end{aligned} \tag{5.40}$$

The equation (5.39) includes $(n-1)$ elimination steps. Since A is sparse, the permutation matrices are chosen such that L and U are sparse too. For the reason of finite storage capacity only nonzero elements are stored. In this context further storage techniques were dealt with previously in Chapter 2. In order to preserve sparseness in the Gauss decomposition monitoring of stability is of course necessary.

Another possible way of solving (5.38) is by partitioning into block matrices. For this reason the following system of linear equations is considered:

$$Ax = b; \quad A := [a_{ik}], \quad A \in \mathbb{R}^{n \times n}, \quad x, b \in \mathbb{R}^n, \tag{5.41}$$

where A is symmetric and positive definite.

Now the system (5.41) is partitioned into block matrices $A_{ik} \neq 0$ of the following form:

$$\begin{bmatrix} A_{11} & A_{21}^T & & & \\ A_{21} & A_{22} & A_{32}^T & & 0 \\ & A_{32} & A_{33} & A_{43}^T & \\ & 0 & & \ddots & \\ & & & & A_{mm} \end{bmatrix} \cdot \begin{bmatrix} x^1 \\ x^2 \\ x^3 \\ \vdots \\ x^m \end{bmatrix} = \begin{bmatrix} b^1 \\ b^2 \\ b^3 \\ \vdots \\ b^m \end{bmatrix}. \tag{5.42}$$

Then A is a is a tridiagonal block matrix with block matrices $A_{ik} \in \mathbb{R}^{r \times r}$ and partial vectors $x^i, b^i \in \mathbb{R}^r$, $i = 1, 2, \ldots, m$. Since A is positive definite and symmetric $(A = A^T)$ the diagonal block matrices A_{ii} are symmetric too, i.e. $A_{ii} = A_{ii}^T$ and A_{ii} is positive definite.

So for the 1st block system

$$A_{11} x^1 + A_{21}^T x^2 = b^1$$

follows. According to the assumption, A_{11}^{-1} exists and hence

$$x^1 + C_{21}^T x^2 = d^1, \tag{5.43}$$

where

$$C_{21}^T := A_{11}^{-1} A_{21}^T \quad \text{and} \quad d^1 := A_{11}^{-1} b^1.$$

With the help of (5.43) x^1 is eliminated from the 2nd block system and consequently

$$\hat{A}_{22} x^2 + A_{32}^T x^3 = \hat{b}^2, \tag{5.44}$$

where

$$\hat{A}_{22} := A_{22} - A_{21} C_{21}^T, \quad \hat{b}^2 := b^2 - A_{21} d^1.$$

At this point $C_{21}^T, \hat{A}_{22}, d^1, \hat{b}^2$ are efficiently determinable and the following system is obtained:

$$\begin{bmatrix} I & C_{21}^T & & & \\ & \hat{A}_{22} & A_{32}^T & & 0 \\ & A_{32} & A_{33} & A_{43}^T & \\ & 0 & & \ddots & \\ & & & & A_{mm} \end{bmatrix} \cdot \begin{bmatrix} x^1 \\ x^2 \\ x^3 \\ \vdots \\ x^m \end{bmatrix} = \begin{bmatrix} d^1 \\ \hat{b}^2 \\ b^3 \\ \vdots \\ b^m \end{bmatrix}. \tag{5.45}$$

5.3 Direct algorithms with block matrices

This elimination process may be continued recursively, and after m steps

$$\begin{bmatrix} I & C_{21}^T & & & \\ & I & C_{32}^T & & 0 \\ & & I & C_{43}^T & \\ & 0 & & \ddots & \\ & & & & I \end{bmatrix} \cdot \begin{bmatrix} x^1 \\ x^2 \\ x^3 \\ \vdots \\ x^m \end{bmatrix} = \begin{bmatrix} d^1 \\ d^2 \\ d^3 \\ \vdots \\ d^m \end{bmatrix} \qquad (5.46)$$

is obtained. So the partial vector sought is

$$x^m := d^m.$$

The remaining partial vectors are determined in reverse order by

$$x^i := d^i - C_{i+1,i}^T x^{i+1}, \quad i = m-1, \ldots, 1. \qquad (5.47)$$

Note 5.48
In order to avoid the direct calculation of the inverse $\hat{A}_{ii}^{-1}$ in (5.43) a certain number of linear equation systems with respectively different right-hand sides are solved for determining the block matrices $C_{i+1,i}^T$ in (5.46).
Since $\det \hat{A}_{ii} \neq 0$,

$$\left. \begin{array}{rcl} \hat{A}_{ii} C_{i+1,i}^T & = & A_{i+1,i}^T \\ \hat{A}_{ii} d^i & = & \hat{b}^i \end{array} \right\} \text{ for } i = 1, 2, \ldots, m-1, \qquad (5.49)$$

where

$$\hat{A}_{11} = A_{11} \quad \text{and} \quad \hat{b}^1 = b^1.$$

Here for large m and r the number of operations is significantly reduced against the normal calculation of the inverse.

Number of operations to determine the inverse: $\sim 3(m-2)r^3$.
Number of equation systems: $\sim \frac{5}{3}(m-1)r^3$.

An alternative method is the application of the Cholesky decomposition to the individual block systems. The decomposition

$$A_{11} := L_{11} L_{11}^T$$

and

$$A_{11} x^1 + A_{21}^T x^2 = b^1$$

yields

$$L_{11}^T x^1 + L_{21}^T x^2 = h^1. \qquad (5.50)$$

Thus, the equations

$$\begin{array}{rcl} L_{11} L_{21}^T & = & A_{21}^T \\ L_{11} h^1 & = & b^1 \end{array} \qquad (5.51)$$

lead to L_{21}^T and h^1 by forward substitution.

Since $A_{21} = L_{21}L_{11}^T$ (see (5.50), (5.51)) yields

$$\hat{A}_{22}x^2 + A_{32}^T x^3 = \hat{b}^2, \qquad (5.52)$$

where

$$\hat{A}_{22} := A_{22} - L_{21}L_{21}^T, \quad \hat{b}^2 = b^2 - L_{21}h^1$$

by multiplication by L_{21} and subtraction from the 2nd block equation analogous to (5.44).

After further reduction, the system

$$Lx = h \Leftrightarrow \begin{bmatrix} L_{11}^T & L_{21}^T & & & & \\ & L_{22}^T & L_{32}^T & & 0 & \\ & & L_{33}^T & L_{43}^T & & \\ & 0 & & \ddots & & \\ & & & & & L_{mm}^T \end{bmatrix} \cdot \begin{bmatrix} x^1 \\ x^2 \\ x^3 \\ \vdots \\ x^m \end{bmatrix} = \begin{bmatrix} h^1 \\ h^2 \\ h^3 \\ \vdots \\ h^m \end{bmatrix} \qquad (5.53)$$

is obtained. The coefficient matrix L is a block matrix decomposition of the given matrix A (Cholesky decomposition). At this point of the algorithm the partial vector x^m follows from

$$L_{mm}^T x^m = h^m. \qquad (5.54)$$

The remaining partial vectors are obtained recursively from

$$L_{ii}^T x^i = h^i - L_{i+1,i}^T x^{i+1}, \quad i = m-1, \ldots, 1. \qquad (5.55)$$

Note 5.56
Here for large m, r the number of operations is reduced to $\sim \frac{7}{6}(m-1)r^3$ compared with the process treated above. A generalization to nonsymmetric matrices is possible.

A variant of the direct block process can be applied to tridiagonal block matrices. For this purpose the system of linear equations

$$Ax = b; \quad A := [a_{ik}], \quad A \in \mathbb{R}^{n \times n}, \ x, b \in \mathbb{R}^n, \\ A \text{ nonsingular}, \qquad (5.57)$$

partitioned into block matrices of the following form, is considered:

$$\begin{bmatrix} A_1 & C_1 & & & & \\ B_2 & A_2 & C_2 & & 0 & \\ & B_3 & A_3 & C_3 & & \\ & & \ddots & \ddots & \ddots & \\ & 0 & & B_{m-1} & A_{m-1} & C_{m-1} \\ & & & & B_m & A_m \end{bmatrix} \cdot \begin{bmatrix} x^1 \\ x^2 \\ x^3 \\ \vdots \\ x^m \end{bmatrix} = \begin{bmatrix} b^1 \\ b^2 \\ b^3 \\ \vdots \\ b^m \end{bmatrix}. \qquad (5.58)$$

A is a tridiagonal block matrix, where

$$\begin{aligned} A_i &\in \mathbb{R}^{r_i \times r_i}, & i &= 1, 2, \ldots, m & &\text{(square matrices)} \\ B_i &\in \mathbb{R}^{r_i \times r_{i-1}}, & i &= 2, \ldots, m & &\text{(rectangular matrices)} \\ C_i &\in \mathbb{R}^{r_i \times r_{i+1}}, & i &= 1, \ldots, m-1 & &\text{in general}). \end{aligned} \qquad (5.59)$$

5.3 Direct algorithms with block matrices

If $r_i = r$, $i = 1, 2, \ldots, m$, then all blocks A_i, B_i, C_i are square matrices of order r (see previous process). Then the matrix A in (5.58) is of band structure with band width given by A_i, B_i, C_i.

In the special case $A_i, B_i, C_i \in \mathbb{R}^{1 \times 1}$ the matrix A in (5.58) is a tridiagonal matrix with band width 3.

For the partial vectors $x^i, b^i \in \mathbb{R}^{r_i}$, $i = 1, 2, \ldots, m$,

$$x^i := [x_1^i, \ldots, x_{r_i}^i]^T, \quad i = 1, 2, \ldots, m$$

and

$$b^i := [b_1^i, \ldots, b_{r_i}^i]^T, \quad i = 1, 2, \ldots, m.$$

Now the system $Ax = b$ is transformed into an upper block triangular system

$$Cx = d, \tag{5.60}$$

where C, d are assumed to be

$$C := \begin{bmatrix} F_1 & C_1 & & \\ & F_2 & \ddots & 0 \\ 0 & & \ddots & C_{m-1} \\ & & & F_m \end{bmatrix}, \quad d := \begin{bmatrix} d^1 \\ d^2 \\ \vdots \\ d^m \end{bmatrix}. \tag{5.61}$$

For the new blocks F_i and d^i respectively, the following algorithm is obtained recursively:

$$\left. \begin{aligned} F_1 &:= A_1 \\ d^1 &:= b^1 \\ F_i &:= A_i - B_i F_{i-1}^{-1} C_{i-1} \\ d^i &:= b^i - B_i F_{i-1}^{-1} d^{i-1} \end{aligned} \right\} \quad i = 2, 3, \ldots, m. \tag{5.62}$$

According the assumption $det(A) \neq 0$ and

$$det(A) = det(C) = \prod_{i=1}^{m} det(F_i) \neq 0. \tag{5.63}$$

Hence the diagonal blocks F_i are nonsingular; consequently the algorithm may be carried out without pivoting. In practice the order of the diagonal blocks is generally small.

For calculating the inverse F_i^{-1} special numerically efficient algorithms are used. In order to avoid the direct calculation of the inverse, Gaussian elimination, for example, is used for solving the equation systems

$$D_i := B_i F f_{i-1}^{-1} \quad \Leftrightarrow \quad F_{i-1}^T D_i^T = B_i^T. \tag{5.64}$$

Here D_i^T is the solution matrix of the system equations (5.64).

So by the algorithm (5.62) the system $Cx = d$ is obtained; finally the following partial systems have to be solved:

$$\begin{aligned} F_m x^m &= d^m \\ F_i x^i &= d^i - C_i x^{i+1}, \quad i = m-1, \ldots, 1. \end{aligned} \tag{5.65}$$

The partial systems (5.65) are generally solved by LU decomposition or, in the case of $F_i = F_i^T$, by Cholesky decomposition:

$$\begin{aligned}(L_m U_m)x^m &= d^m \\ (L_i U_i)x^i &= f^i = d^i - C_i x^{i+1}, \quad i = m-1,\ldots,1.\end{aligned} \qquad (5.66)$$

Then the corresponding triangular systems are solved.

Note 5.67
Partial pivoting or complete pivoting in general lead to a reduction of rounding errors and instabilities (cancellation effect). Correspondingly carefully chosen pivot elements were investigated in [8]. As is known, the direct algorithms treated above yield only an approximate solution $\bar{x}$ for x, which, however, can be improved by reiteration.

Let $x^{(0)}$ be an approximate solution of the system $Ax = b$ obtained by the algorithm above; then

$$Ax^{(0)} - b = r^{(0)}. \qquad (5.68)$$

The equation

$$A\delta z^{(0)} = r^{(0)} \qquad (5.69)$$

is then solved by LU factorization, for example, and a refined solution is

$$x^{(1)} := x^{(0)} - \delta z^{(0)}. \qquad (5.70)$$

If $\delta z^{(0)}$ is not small enough, the matrix A is poorly conditioned. Again $x^{(1)}$ is an approximate solution so that the reiteration may be continued according to the following algorithm:

$$\left.\begin{aligned}(i) \quad & Ax^{(i)} - b := r^{(i)} \\ (ii) \quad & Az^{(i)} := r^{(i)} \\ (iii) \quad & x^{(i+1)} := x^{(i)} - z^{(i)}\end{aligned}\right\} \quad i = 0, 1, \ldots . \qquad (5.71)$$

In order that the reiteration be meaningful and lead to a refinement, this algorithm must be applied to the problem very carefully in the light of the numerical accuracy of the system being used.

Note 5.72
The question arises whether, for example, direct or iterative algorithms are preferable for the numerical treatment of elliptic partial differential equations. In the case of finite differences and simple finite elements the matrices are band matrices and very sparse so that iterative algorithms are appropriate. For large and complicated finite elements the matrices are often less sparse. In this case direct methods are likely to be preferable.

5.4 Least squares problems

The overdetermined system of linear equations

$$Ax = b; \quad A := [a_{ik}], \quad A \in \mathbb{R}^{m \times n}, \; x \in \mathbb{R}^n, \; b \in \mathbb{R}^m \tag{5.73}$$

is considered, where A is sparse, $rank(A) = n$ and $m > n$.

Since there is in general no exact solution for an overdetermined system, the minimization of the error can be tried, i.e. the minimum of

$$f(x) = \|Ax - b\|_p$$

for an appropriate norm is sought.

This means for the method of least squares:

$$\min_x \|Ax - b\|_2, \quad A \in \mathbb{R}^{m \times n}, \; b \in \mathbb{R}^m. \tag{5.74}$$

By making use of an orthogonal matrix Q, $Q^T Q = I_m$, (5.74) can be transformed into the equivalent problem

$$\min_x \|(Q^T A)x - Q^T b\|_2. \tag{5.75}$$

Method of least squares

Let the overdetermined system of m linear equations

$$\sum_{k=1}^{n} a_{ik} x_k + b_i = r_i, \quad i = 1, 2, \ldots, m, \quad n < m, \tag{5.76}$$

be given with n unknown $x_1, x_2, \ldots x_n$, i.e. in matrix notation the system of *error equations*

$$Ax + b = r; \quad A \in \mathbb{R}^{m \times n}, \; x \in \mathbb{R}^n, \; b, r \in \mathbb{R}^m \tag{5.77}$$

is assumed given. Furthermore, the matrix A is assumed to have the maximal rank with $rank(A) = n$. r is called the *residual vector*.

According to the Gaussian adjustment principle the unknowns x_k of the error equations are determined in such a way that the sum of the squares of the residues r_i is minimal. This condition is equivalent to the minimization of $\|r\|_2^2$. Equation (5.77) implies

$$\begin{aligned} r^T r &= (Ax + b)^T (Ax + b) \\ &= x^T A^T A x + x^T A^T b + b^T A x + b^T b \\ &= x^T A^T A x + 2(A^T b)^T x + b^T b. \end{aligned} \tag{5.78}$$

$r^T r$ is representable as a quadratic $\mathcal{F}(x)$ in the unknown $x^T = [x_1, \ldots, x_n]$.

By the first Gauss transformation
$$C := A^T A; \quad C := [c_{ik}], \quad C \in \mathbb{R}^{n \times n},$$
$$C \text{ symmetric and positive definite} \quad (5.79)$$
$$d := A^T b; \quad d \in \mathbb{R}^n$$

the *minimization problem* for $\mathcal{F}(x)$ is:
$$\mathcal{F}(x) := r^T r = x^T C x + 2 d^T x + b^T b = \min_x. \quad (5.80)$$

A necessary condition for $\mathcal{F}(x)$ to be minimized is
$$\nabla \mathcal{F}(x) = 0; \quad (\nabla : \text{gradient of } \mathcal{F}(x)),$$
which means that the following system of equations has to be solved:
$$Cx + d = 0; \quad (\text{normal form of the error equations}). \quad (5.81)$$

The ith component of $\nabla \mathcal{F}(x)$ in (5.80) is:
$$\frac{\partial \mathcal{F}(x)}{\partial x_i} = 2 \sum_{k=1}^{n} c_{ik} x_k + 2 d_i, \quad i = 1, 2, \ldots, n. \quad (5.82)$$

Since C is positive definite, according to the assumption, the unknown v_i, $i = 1, 2, \ldots, n$, of the normal form (5.81) are uniquely determinable.

Various solution methods

Classical method

Let a_i be the column vector of A; then c_{ik} and d_i can be determined by
$$c_{ik} := a_i^T a_k$$
$$d_i := a_i^T b \quad i, k = 1, 2, \ldots, n,$$

and the following algorithm for solving the problem will be obtained:

$$(i) \quad \left. \begin{array}{rcl} C & := & A^T A \\ d & := & A^T b \end{array} \right\} \text{Gauss transformation, normal form}$$

$$(ii) \quad \left. \begin{array}{rcl} C & := & LL^T \\ Ly & = & d \\ L^T x & = & -y \end{array} \right\} \begin{array}{l} \text{Cholesky decomposition} \\ \text{substitution in the triangular systems} \end{array} \quad (5.83)$$

$$(iii) \quad r = Ax + b \quad \text{determination of the residues}$$

If the matrix C of the normal form is of poor condition, then the relative error of the solution $\tilde{x}$ is large. So, in such cases numerically more stable equations have to be considered, and these are obtained by orthogonal transformations.

Orthogonal transformation method

Since the length of a vector is invariant under orthogonal transformation, the error equations (5.77) may be transformed by an orthogonal matrix $Q \in \mathbb{R}^{m \times m}$ without altering the squares of the residues.

5.4 Least squares problems

A generalization of the Theorem concerning the QR decomposition of a matrix is the following

Theorem 5.84
For each matrix $A \in I\!R^{m \times n}$ having maximal rank $n < m$ there exists an orthogonal matrix $Q \in I\!R^{m \times m}$ with

$$A = Q\hat{R}, \text{ where } \hat{R} = \begin{bmatrix} R \\ 0 \end{bmatrix}, R \in I\!R^{n \times n}, 0 \in I\!R^{(m-n) \times n}$$

and R is a *regular* upper triangular matrix.

Now the following system, equivalent to (5.77), will be considered:

$$Q^T A x + Q^T b = Q^T r = \hat{r}, \text{ where } Q^T Q = I_m. \tag{5.85}$$

Theorem 5.84 and equation (5.85) imply

$$\hat{R} x + \hat{b} = \hat{r}; \quad \hat{R} = Q^T A, \ \hat{b} = Q^T b,$$

i.e.

$$\left. \begin{array}{rcl} r_{11} x_1 + r_{12} x_2 + \cdots + r_{1n} x_n + \hat{b}_1 &=& \hat{r}_1 \\ r_{22} x_2 + \cdots + r_{2n} x_n + \hat{b}_2 &=& \hat{r}_2 \\ & \vdots & \\ r_{nn} x_n + \hat{b}_n &=& \hat{r}_n \end{array} \right\} n$$

$$\left. \begin{array}{rcl} \hat{b}_{n+1} &=& \hat{r}_{n+1} \\ & \vdots & \\ \hat{b}_m &=& \hat{r}_m \end{array} \right\} m - n. \tag{5.86}$$

Since the $(m-n)$ residues $\hat{r}_{n+1}, \ldots, \hat{r}_m$ are determined by the corresponding $\hat{b}_{n+1}, \ldots, \hat{b}_m$ and do not depend on x_k, $\sum_{i=1}^m \hat{r}_i^2$ is minimal, if

$$\hat{r}_1 = \hat{r}_2 = \ldots = \hat{r}_n = 0;$$

consequently

$$\sum_{i=1}^m \hat{r}_i^2 = \sum_{k=n+1}^m \hat{r}_k^2.$$

Hence the unknown x_k, $k = 1, \ldots, n$, are determined by

$$R x + \hat{b}_e := 0, \ \hat{b}_e \in I\!R^n, \tag{5.87}$$

where $\hat{b}_e$ comprises the initial n components of $\hat{b} = Q^T b$,

$$\hat{b} := \begin{bmatrix} \hat{b}_e \\ \hat{r}_{n+1} \\ \vdots \\ \hat{r}_m \end{bmatrix} \begin{array}{l} \}n \\ \}m-n. \end{array}$$

By backward substitution (5.87) yields x. Then the complete algorithm is:

$$\begin{aligned}(i) \quad & A = Q\hat{R} && QR \text{ decomposition} \\ (ii) \quad & \hat{b} = Q^T b && Q\text{-transformation of } b \\ (iii) \quad & Rx + \hat{b}_e = 0 && \text{backward substitution} \\ (iv) \quad & r = Q\hat{r} && \text{backward substitution of } \hat{r}.\end{aligned} \quad (5.88)$$

The algorithm (5.88) is more efficient and more stable than algorithm (5.83).

Note 5.89
If there is only one system of error equations to be solved, (i) and (ii) of the algorithm (5.88) are generally calculated simultaneously. As soon as several systems must be solved having the same matrix A, but different elements b in the equation $Ax + b = r$, a separation of the steps (i) and (ii) is necessary. In this case parallel computing can possibly lead to an improvement.

Solution by using Householder transformations
Once again the system of error equations (5.77)

$$Ax + b = r$$

is considered. The orthogonal transformation is achieved by means of the Householder transformation

$$H := I - 2ww^T \quad \text{where } w^T w = 1; \ w \in \mathbb{R}^m, \ H \in \mathbb{R}^{m \times m}. \quad (5.90)$$

Then H is symmetric ($H = H^T$) and orthogonal ($H^T H = I$).
After n transformations

$$\begin{aligned} H_n H_{n-1} \ldots H_2 H_1 A &= \hat{R} \\ H_n H_{n-1} \ldots H_2 H_1 b &= \hat{b} \end{aligned} \quad (5.91)$$

is obtained.
The solution x follows from

$$Rx + \hat{b}_e = 0$$

by backward substitution. For the residual vector r

$$H_n H_{n-1} \ldots H_2 H_1 \cdot r = \hat{r} = [0, 0, \ldots, 0, b_{n+1}, \hat{b}_{n+1}, \ldots, \hat{b}_m]. \quad (5.92)$$

Taking into account $H_i = H_i^T$ the equation

$$r = H_1 H_2 \ldots H_{n-1} H_n \hat{r} \quad (5.93)$$

will follow.

5.4 Least squares problems

Singular value decomposition

The preceding orthogonal transformations were assumed to have maximal rank n, i.e. $rank(A) = n$, $n < m$. In applied mathematics, however, this does not always occur. So more general decompositions must be sought. For this purpose the following results are quoted:

Theorem 5.94
For each matrix $A \in {\rm I\!R}^{m \times n}$ having $rank(A) = r$, $r < n < m$, there exist 2 orthogonal transformations $Q \in {\rm I\!R}^{m \times m}$ and $U \in {\rm I\!R}^{n \times n}$ so that

$$Q^T A U = \hat{R} \quad \text{where} \quad \hat{R} = \begin{bmatrix} R & 0_1 \\ 0_2 & 0_3 \end{bmatrix}, \quad \hat{R} \in {\rm I\!R}^{m \times n}, \, R \in {\rm I\!R}^{r \times r}. \tag{5.95}$$

Here R is a nonsingular upper triangular matrix. The matrices 0_i, $i = 1, 2, 3$, are zero matrices of appropriate dimensions.

Theorem 5.96
For each matrix $A \in {\rm I\!R}^{m \times n}$ having $rank(A) = r$, $r \leq n < m$, there exist 2 orthogonal transformations $U \in {\rm I\!R}^{m \times m}$ and $V \in {\rm I\!R}^{n \times n}$ so that the following *singular value decomposition*

$$A = U \hat{S} V^T, \quad \text{where} \quad \hat{S} = \begin{bmatrix} S \\ 0 \end{bmatrix}, \quad \hat{S} \in {\rm I\!R}^{m \times n}, \, S \in {\rm I\!R}^{n \times n}, \tag{5.97}$$

holds. Here

$$S = \begin{bmatrix} s_1 & & & 0 \\ & s_2 & & \\ & & \ddots & \\ 0 & & & s_n \end{bmatrix} \quad \text{is a diagonal matrix,}$$

where $s_1 \geq s_2 \geq \ldots \geq s_r > s_{r+1} = s_{r+2} = \ldots = s_n = 0$ and 0 is a zero matrix. The elements s_i, $i = r+1, \ldots, n$ are called *singular values* of the matrix A.

Let $u_i \in {\rm I\!R}^m$ and $v_i \in {\rm I\!R}^n$ be the column vectors of U and V respectively. Then (5.97) implies

$$\begin{array}{rcl} A v_i & = & s_i u_i \\ A^T u_i & = & s_i v_i \end{array} \quad i = 1, 2, \ldots, n. \tag{5.98}$$

The vectors v_i are called *right-hand singular vectors* and the vectors u_i *left-hand singular vectors* of A.

Such a decomposition allows the system of normal equations to be transformed into an equivalent system by the above-mentioned orthogonal transformation. The condition $V^T V = I$ yields

$$U^T A V V^T x + U^T b = U^T r = r. \tag{5.99}$$

By means of the auxiliary vectors

$$y := V^T x, \quad y \in \mathbb{R}^n$$
$$c := U^T b, \quad c \in \mathbb{R}^m, \quad \text{where} \quad c_i := U_i^T b \qquad (5.100)$$

and the decompositions

$$U^T A V = S \quad \text{and} \quad U U^T A V V^T = U S V^T, \qquad (5.101)$$

equation (5.99) is transformed into

$$\begin{aligned} s_i y_i + c_i &= \hat{r}_i, \quad i = 1, 2, \ldots, r \\ c_i &= \hat{r}_i, \quad i = r+1, \ldots, m. \end{aligned} \qquad (5.102)$$

The vectors $\hat{r}_i$ do not depend on the unknown y_i. Hence $\sum_{i=1}^m \hat{r}_i^2$ is minimal, if $\hat{r}_i = 0$, $i = 1, 2, \ldots, r$, and the minimum is

$$\min_i := r^T r = \sum_{i=r+1}^m \hat{r}_i^2 = \sum_{i=r+1}^m c_i^2 = \sum_{i=r+1}^m (U_i^T b)^2. \qquad (5.103)$$

The initial r unknowns y_i are given by

$$y_i := -\frac{c_i}{s_i}, \quad i = 1, 2, \ldots, r. \qquad (5.104)$$

The remaining $(n-r)$ variables may be chosen freely. So the solution vector x of the normal equations of (5.100) is

$$x := -\sum_{i=1}^r \frac{u_i^T b}{s_i} v_i + \sum_{i=r+1}^n y_i v_i \qquad (5.105)$$

with $(n-r)$ free parameters $y_{r+1}, y_{r+2}, \ldots, y_n$ [20].

If the rank of the matrix A is not maximal, i.e. $rank(A) < n$, then, as is known, the general solution is the sum of an arbitrary vector in the zero space of the matrix A and a special solution in the subspace of the r right-hand singular vectors v_i corresponding to the positive singular values s_i.

For the singular values $s_i := 0$, $i = r+1, \ldots, n$, the equation $A v_i = s_i u_i$ implies that

$$A v_i = 0, \quad i = r+1, \ldots, n.$$

The set of solutions of the system of normal equations contains a special solution x^*, minimal in the Euclidian norm. Since the vectors v_i are orthogonal and $y_{r+1} = y_{r+2} = y_n = 0$ the special solution is

$$x^* = -\sum_{i=1}^r \frac{u_i^T b}{s_i} v_i, \quad \|x^*\|_2 = \min_{Ax+b=r} \|x\|_2. \qquad (5.106)$$

The decomposition of A gives insight into the construction of the general solution and the special solution x^*.

6 Iterative algorithms

The problem with which we begin the discussion is again the determination of a unique solution of the system

$$Ax = b, \quad \text{where } A := [a_{ik}], \quad A \in \mathbb{R}^{n \times n} \text{ or } A \in \mathbb{C}^{n \times n}, \quad A \text{ sparse.} \qquad (6.1)$$

In the context of sparse systems, iterative algorithms possess a unique advantage when the structure of A is not clearly recognizable. Iterative algorithms are then more effective in terms of the use of the store and the number of mathematical operations since only the elements $a_{ik} \neq 0$ are used.

Principally, iterative algorithms should

- preserve the original sparseness,

- minimize the fill-in.

The particular structure of the matrix A is irrelevant.

A disadvantage of these iterative algorithms is their slow convergence. Acceleration of the convergence is achieved by means of the residual vector

$$r^{(k)} := b - Ax^{(k)}, \quad k = 0, 1, 2, \ldots . \qquad (6.2)$$

Extrapolation techniques are constructed by $r^{(k)}$:

$$x^{(k+1)} := x^{(k)} + q \cdot r^{(k)}, \quad k = 0, 1, 2, \ldots, \quad x^{(0)} \in \mathbb{R}^n, \qquad (6.3)$$

where $q := q(\max_i(\lambda_i))$, and λ_i is an eigenvalue of A.

Simultaneous displacement method:

$$\left.\begin{array}{rl} x^{(k+1)} & := x^{(k)} + qD^{-1}(b - Ax^{(k)}) \\ & = (I - qD^{-1}A)x^{(k)} + qD^{-1}b \end{array}\right\} k = 0, 1, 2, \ldots, \quad x^{(0)} \in \mathbb{R}^n \quad (6.4)$$

Richardson method:

$$\left.\begin{array}{rl} x^{(k+1)} & := x^{(k)} + q_k D^{-1} r^{(k)} \\ & = (I - q_k D^{-1} A)x^{(k)} + q_k D^{-1} b \end{array}\right\} k = 0, 1, 2, \ldots, \quad x^{(0)} \in \mathbb{R}^n \quad (6.5)$$

where $q_k = q_k(\max_i(\lambda_i))$.

If $a < \lambda_i < b$ and $p = b/a$ is the *condition number* of the matrix A, the *rate of convergence* of the displacement method, or the Richardson method, is $\frac{2}{p}$ or $\frac{2}{\sqrt{p}}$, respectively.

If there is still some storage capacity available, the iterates are incorporated into the algorithm:

$$\begin{array}{rl} Dx^{(k+1)} & := Dx^{(k)} + q(b - Ax^{(k)}) + pD(x^{(k)} - x^{(k-1)}) \\ Dx^{(k+1)} & := Dx^{(k)} + q_k(b - Ax^{(k)}) + p_k D(x^{(k)} - x^{(k-1)}). \end{array} \quad (6.6)$$

p, q, p_k, q_k are acceleration parameters for the convergence.

6.1 Some properties of sparse matrices

We now summarize some interesting properties of sparse matrices which have been already touched on in a different context.

Definition 6.7

A complex matrix $A := [a_{ik}]$, $A \in \mathbb{C}^{n \times n}$, is called *weakly diagonally dominant* if

$$|a_{ii}| \geq \sum_{\substack{k=1 \\ k \neq i}}^{n} |a_{ik}| \quad \text{for all } i = 1, 2, \ldots, n \quad (6.8)$$

and

$$|a_{ii}| > \sum_{\substack{k=1 \\ k \neq i}}^{n} |a_{ik}| \quad \text{for at least one } i. \quad (6.9)$$

This property is also called *the weak row sum criterion*.

Definition 6.10

A complex matrix $A := [a_{ik}]$, $A \in \mathbb{C}^{n \times n}$, is called *reducible*, or *decomposable*, if $n > 1$ and if there are non-empty subsets S and T of the set $W = \{1, 2, \ldots, n\}$ having the following properties:

$$\begin{array}{rl} (i) & S \cap T = \emptyset \\ (ii) & S \cup T = W \\ (iii) & a_{ik} = 0 \quad \text{for each } i \in S \text{ and each } k \in T. \end{array}$$

A is said to be *irreducible* or *non-decomposable* if for each partition of W into S, and T satisfying $(i), (ii), (iii)$, there is at least one element $a_{ik} \neq 0$, where $i \in S$ and $k \in T$.

Example 6.11

(i) Trivially every matrix having no zero elements is irreducible.

(ii) The matrix $A := \begin{bmatrix} 1 & 0 & 1 \\ 0 & 1 & 0 \\ 1 & 1 & 1 \end{bmatrix}$ is reducible ($S = \{2\}$ and $T = \{1, 3\}$).

(iii) The matrix $B := \begin{bmatrix} 1 & 0 & 1 \\ 0 & 1 & 1 \\ 1 & 1 & 0 \end{bmatrix}$ is, however, irreducible.

Theorem 6.12
The matrix $A := [a_{ik}]$, $i, k = 1, 2, \ldots, n$, is *irreducible* if, and only if, $n = 1$ or if $n > 1$ and for each pair $(i, k) \in \{1, 2, \ldots, n\} \times \{1, 2, \ldots, n\}$, where $i \neq k$, either $a_{ik} \neq 0$ or there are indices $i_1, i_2, \ldots, i_r$, such that

$$a_{ii_1} \cdot a_{i_1 i_2} \cdot \ldots \cdot a_{i_r i} \neq 0. \tag{6.13}$$

Proof:
By definition A is irreducible if $n = 1$.

(i) Let $W = \{1, 2, \ldots, n\}$, $S \subset W$, $T \subset W$, $S \neq \emptyset$, $T \neq \emptyset$, $S \cap T \neq \emptyset$, and $S \cup T = W$.

Let $n > 1$ and let condition (6.13) hold for $i, k \in W$, $i \neq k$; also let $i \in S$ and let $k \in T$ be arbitrary. Then either $a_{ik} \neq 0$ or there are indices $i_1, i_2, \ldots, i_r$ such that $a_{ii_1} \cdot a_{i_1 i_2} \cdot \ldots \cdot a_{i_r k} \neq 0$. At least one of the factors has its first index in S and its second index in T.

One commences with the first factor. $i \in S$ and $i_1 \in T$ already imply irreducibility. If $i_1 \notin T$, then $i_1 \in S$ and we next look at the second factor $a_{i_1 i_2}$ with $i_1 \in S$. If $i_2 \in T$, the assertion again follows. If $i_2 \notin T$, then $i_2 \in S$ and so we pass to the third factor.

This process may be continued until either a factor is found whose second index is in T or one arrives at the final factor whose first index is in S and whose second index is in T, according to the assumption. S and T were assumed to be an arbitrary partition of W. This implies that for each partition there exists an element $a_{ik} \neq 0$ whose first index is in S and whose second index is in T and thus irreducibility will follow.

(ii) Next we assume that $n > 1$ and that A is irreducible.

Let $i, k \in W$, $i \neq k$, and $a_{ik} = 0$. Then let $S_1 = \{i\}$ and $T_1 = W \setminus S_1$. Irreducibility then implies the existence of an element $i_1 \in T_1$ with $a_{ii_1} \neq 0$.

Furthermore let $S_2 = \{i, i_1\}$ and $T_2 = W \setminus S_2$, where $k \neq i_1$ and therefore $T_2 \neq \emptyset$. If $a_{i_1 k} \neq 0$, then (6.13) holds with $r = 1$. In the other case let $i_2 \in T_2$ so that $a_{ii_2} \neq 0$ or $a_{i_1 i_2} \neq 0$ (i_2 exists because of the irreducibility).

Now let $S_3 = \{i, i_1, i_2\}$ and $T_3 = W \setminus S_3$. Evidently $k \neq i_2$ and therefore $T_3 \neq \emptyset$. If $a_{i_2 k} \neq 0$, then (6.13) holds with $r = 2$. Otherwise we continue this process until (6.13) can be satisfied for certain elements r.

Therewith Theorem 6.12 is proved. $\square$

To demonstrate the existence of a unique solution of $Ax = b$ it must be shown that $det(A) \neq 0$. For this purpose we propose the following Theorem:

Theorem 6.14
If A is an irreducible matrix of order n and A is weakly diagonally dominant, then $det(A) \neq 0$.

Proof:
First we shall show that $a_{ii} \neq 0$ for all i:

If $n = 1$, then $|a_{11}| > 0$; hence $a_{11} \neq 0$. This follows from the weak diagonal dominance of A. It then follows that $det(A) \neq 0$.

For $n > 1$ we assume $a_{ii} = 0$ for a special i. Then because of the weak diagonal dominance $a_{ik} = 0$ for all k. Therefore there are no $i_1, \ldots, i_r$ such that (6.13) holds for $k \neq i$. Therefore, following Theorem 6.12, the matrix A is not irreducible. This contradiction thus shows that $a_{ii} \neq 0$ for all i.

If $det(A) = 0$, then there exists a nontrivial solution u of the homogeneous system

$$Au = 0. \tag{6.15}$$

Since $a_{ii} \neq 0$ for all i, the ith equation is solvable for u_i and one obtains

$$u_i := \sum_{k=1}^{n} b_{ik} u_k, \quad i = 1, 2, \ldots, n, \tag{6.16}$$

where

$$b_{ii} := 0 \quad \text{for} \quad i = k \quad \text{and} \quad b_{ik} := -\frac{a_{ik}}{a_{ii}} \quad \text{for} \quad i \neq k. \tag{6.17}$$

Because of the weak diagonal dominance

$$\sum_{k=1}^{n} |b_{ik}| \leq 1 \tag{6.18}$$

holds for all i, and for some i the inequality is strict.

Since u is a nontrivial solution of $Au = 0$, there exists a number m defined by

$$m := \max_{i}(|u_i|) > 0.$$

Let l be the index such that $|u_l| = m$. Then (6.16) implies

$$u_l := \sum_{k=1}^{n} b_{lk} u_k \tag{6.19}$$

and from (6.18) we obtain

$$\sum_{k=1}^{n} |b_{lk}| |u_k| \geq |u_l| \geq \sum_{k=1}^{n} |b_{lk}| |u_l| = \sum_{k=1}^{n} |b_{lk}| \cdot m. \tag{6.20}$$

It follows that

$$\sum_{k=1}^{n} |b_{lk}| (|u_k| - |u_l|) \geq 0. \tag{6.21}$$

Since $|u_l| \geq |u_k|$ for all k, the above inequality is satisfied only if
$$|b_{lk}|(|u_k|-|u_l|) = 0 \text{ for all } k.$$
Thus, for k, with $b_{lk} \neq 0$, we have
$$|u_k| = |u_l| = m. \tag{6.22}$$
For $k \neq l$ the irreducibility of A implies the existence of a sequence of indices $i_1, i_2, \ldots, i_r$ such that (6.13) is satisfied for $i = l$. Hence all $b_{li_1}, b_{i_1 i_2}, \ldots, b_{i_r k}$ are different from zero.

Therefore we have
$$m = |u_l| = |u_{i_1}| = |u_{i_2}| = \ldots = |u_k|, \tag{6.23}$$
and this implies that
$$|u_i| = m \text{ for all } i. \tag{6.24}$$
Now let i_0 be a particular value of i such that
$$\sum_{k=1}^{n} |b_{i_0 k}| < 1. \tag{6.25}$$
Then, according to (6.16), we have
$$|u_{i_0}| \leq \sum_{k=1}^{n} |b_{i_0 k}| |u_k|. \tag{6.26}$$
Since $|u_1| = |u_2| = \ldots = |u_n| = m$, the inequality
$$m = |u_{i_0}| \leq \sum_{k=1}^{n} |b_{i_0 k}| m \tag{6.27}$$
or
$$1 \leq \sum_{k=1}^{n} |b_{i_0 k}| \tag{6.28}$$
hold, thus contradicting (6.25). This proves the Theorem. □

Theorem 6.29
Let A be a real symmetric, irreducible and weakly diagonally dominant matrix, whose diagonal elements are non-negative. Then A is positive definite.

Proof:
Since the matrix A is real and symmetric the eigenvalues of A are real.
Assumption: $\lambda \leq 0$ is an eigenvalue of A.
Since A is irreducible, the matrix $A - \lambda I$ is also irreducible. Since $\lambda \leq 0$ and A is weakly diagonally dominant with non-negative diagonal elements, the matrix $A - \lambda I$ is weakly diagonally dominant too.

Then Theorem 6.14 states $det(A - \lambda I) \neq 0$ and thus λ is not an eigenvalue of A. Hence all eigenvalues of A are positive. □

In this context an additional idea will be introduced concerning sparse matrices whose nonzero elements have a certain pattern:

Definition 6.30
A $(n \times n)$ matrix A is said to possess *property* $\mathcal{A}$ if for $\mathcal{W} := \{1, 2, \ldots, n\}$ there exist two disjoint subsets $\mathcal{S}_1 \subset \mathcal{W}, \mathcal{S}_2 \subset \mathcal{W}$, such that $\mathcal{S}_1 \cup \mathcal{S}_2 = \mathcal{W}$ and, if $i \neq k$, and if $a_{ik} \neq 0$ or $a_{ki} \neq 0$, then $i \in \mathcal{S}_1$ and $k \in \mathcal{S}_2$ or $i \in \mathcal{S}_2$ and $k \in \mathcal{S}_1$, respectively.

Observe that either $\mathcal{S}_1$ or $\mathcal{S}_2$ can be empty. In this case A is a diagonal matrix. If neither $\mathcal{S}_1$ nor $\mathcal{S}_2$ is empty, then by rearranging the rows and the corresponding columns of the matrix A possessing property $\mathcal{A}$ the following form of A can be obtained:

$$A := \begin{bmatrix} D_1 & H \\ K & D_2 \end{bmatrix}, \tag{6.31}$$

where D_1 and D_2 are square diagonal matrices. In this context the following Theorems are quoted:

Theorem 6.32
If a matrix A possesses property $\mathcal{A}$, then so also does the matrix $\tilde{A} := P^{-1}AP$, where P is an arbitrary permutation matrix [62].

Theorem 6.33
A matrix A possesses property $\mathcal{A}$ if, and only if, A is a diagonal matrix or there exists a permutation matrix P so that $P^{-1}AP$ is of the form

$$\tilde{A} = P^{-1}AP = \begin{bmatrix} D_1 & H \\ K & D_2 \end{bmatrix}, \tag{6.34}$$

where D_1 and D_2 are square diagonal matrices.

6.2 The Jacobi method

We shall discuss next a very simple method for the solution of sparse systems due to *Jacobi*. To make sure that the algorithm can be used without difficulty it will be assumed that no diagonal element of A vanishes.

By Theorem 6.14 this precondition is fulfilled if A is irreducible and weakly diagonally dominant. The definition of positive definite matrices implies that such a matrix possesses positive diagonal elements. Furthermore, the condition $a_{ii} \neq 0$ is satisfied if A is an L-matrix (see Definition 6.88). We write the system (6.1) in the form

$$x_i = \sum_{k=i}^{n} b_{ik} x_k + c_i \tag{6.35}$$

or

$$x = Bx + c \tag{6.36}$$

where $B := [b_{ik}]$, $i, k = 1, 2, \ldots, n$ and

$$b_{ik} := \begin{cases} -\frac{a_{ik}}{a_{ii}}, & \text{if } i \neq k \\ 0, & \text{if } i = k \end{cases} \tag{6.37}$$

6.2 The Jacobi method

and $c := [c_1, c_2, \ldots, c_n]^T$ with

$$c_i := \frac{b_i}{a_{ii}}, \quad i = 1, 2, \ldots, n. \tag{6.38}$$

It is assumed, naturally, that no a_{ii} vanishes.
For $n = 3$ the following alternative system holds:

$$\begin{array}{rcrcrcl} x_1 &=& & & b_{12}x_2 &+& b_{13}x_3 + c_1 \\ x_2 &=& b_{21}x_1 &+& & & b_{23}x_3 + c_2 \\ x_3 &=& b_{31}x_1 &+& b_{32}x_2 & & + c_3. \end{array} \tag{6.39}$$

We consider the general system $x = Bx + c$ with

$$B = D^{-1}C, \quad c = D^{-1}b, \tag{6.40}$$

with

$$C := D - A, \quad D := diag(A) = [a_{ii}]. \tag{6.41}$$

For $n = 3$ we obtain

$$C = \begin{bmatrix} 0 & -a_{12} & -a_{13} \\ -a_{21} & 0 & -a_{23} \\ -a_{31} & -a_{32} & 0 \end{bmatrix}, \quad D = \begin{bmatrix} a_{11} & 0 & 0 \\ 0 & a_{22} & 0 \\ 0 & 0 & a_{33} \end{bmatrix},$$

$$B = \begin{bmatrix} 0 & -\frac{a_{12}}{a_{11}} & -\frac{a_{13}}{a_{11}} \\ -\frac{a_{21}}{a_{22}} & 0 & -\frac{a_{23}}{a_{22}} \\ -\frac{a_{31}}{a_{33}} & -\frac{a_{32}}{a_{33}} & 0 \end{bmatrix}, \quad c = \begin{bmatrix} \frac{b_1}{a_{11}} \\ \frac{b_2}{a_{22}} \\ \frac{b_3}{a_{33}} \end{bmatrix}$$

$$\tag{6.42}$$

and finally

$$\begin{bmatrix} x_1 \\ x_2 \\ x_3 \end{bmatrix} = \begin{bmatrix} 0 & b_{12} & b_{13} \\ b_{21} & 0 & b_{23} \\ b_{31} & b_{32} & 0 \end{bmatrix} \cdot \begin{bmatrix} x_1 \\ x_2 \\ x_3 \end{bmatrix} + \begin{bmatrix} c_1 \\ c_2 \\ c_3 \end{bmatrix}. \tag{6.43}$$

The Jacobi algorithm begins with arbitrary initial values

$$x_1^{(0)}, x_2^{(0)}, \ldots, x_n^{(0)}, \quad x_i^{(0)} \in \mathbb{R}$$

and then pruduces iterates for $r = 0, 1, 2, \ldots$:

$$x_i^{(r+1)} := \sum_{k=1}^{n} b_{ik} x_k^{(r)} + c_i, \quad i = 1, 2, \ldots, n, \tag{6.44}$$

or

$$x^{(r+1)} := Bx^{(r)} + c, \tag{6.45}$$

respectively.

For $n = 3$ we have

$$\begin{bmatrix} x_1^{(r+1)} \\ x_2^{(r+1)} \\ x_3^{(r+1)} \end{bmatrix} := \begin{bmatrix} 0 & b_{12} & b_{13} \\ b_{21} & 0 & b_{23} \\ b_{31} & b_{31} & 0 \end{bmatrix} \cdot \begin{bmatrix} x_1^{(r)} \\ x_2^{(r)} \\ x_3^{(r)} \end{bmatrix} + \begin{bmatrix} c_1 \\ c_2 \\ c_3 \end{bmatrix}. \quad (6.46)$$

Example 6.47

$$Ax = b \Leftrightarrow \begin{bmatrix} 4 & -1 & -1 & 0 \\ -1 & 4 & 0 & -1 \\ -1 & 0 & 4 & -1 \\ 0 & -1 & -1 & 4 \end{bmatrix} \cdot \begin{bmatrix} x_1 \\ x_2 \\ x_3 \\ x_4 \end{bmatrix} = \begin{bmatrix} 0 \\ 0 \\ 1000 \\ 1000 \end{bmatrix}.$$

For $x = Bx + c$ one obtains:

$$\begin{bmatrix} x_1 \\ x_2 \\ x_3 \\ x_4 \end{bmatrix} = \begin{bmatrix} 0 & \frac{1}{4} & \frac{1}{4} & 0 \\ \frac{1}{4} & 0 & 0 & \frac{1}{4} \\ \frac{1}{4} & 0 & 0 & \frac{1}{4} \\ 0 & \frac{1}{4} & \frac{1}{4} & 0 \end{bmatrix} \cdot \begin{bmatrix} x_1 \\ x_2 \\ x_3 \\ x_4 \end{bmatrix} + \begin{bmatrix} 0 \\ 0 \\ 250 \\ 250 \end{bmatrix}$$

and so, from (6.45),

$$\begin{aligned}
x_1^{(r+1)} &:= & \tfrac{1}{4}x_2^{(r)} + \tfrac{1}{4}x_3^{(r)} & \\
x_2^{(r+1)} &:= \tfrac{1}{4}x_1^{(r)} & & + \tfrac{1}{4}x_4^{(r)} \\
x_3^{(r+1)} &:= \tfrac{1}{4}x_1^{(r)} & & + \tfrac{1}{4}x_4^{(r)} + 250 \\
x_4^{(r+1)} &:= & \tfrac{1}{4}x_2^{(r)} + \tfrac{1}{4}x_3^{(r)} & + 250.
\end{aligned}$$

Let

$$e_i^{(r)} := x_i^{(r)} - x_i$$

be the error at the r-th step, $r = 0, 1, 2, \ldots$. Commencing with

$$x_1^{(0)} = x_2^{(0)} = x_3^{(0)} = x_4^{(0)} = 0$$

the algorithm can be seen to converge rapidly. The error is reduced by the factor $\frac{1}{2}$ after every iteration step and we obtain an error $e_i^{(23)} < 10^{-4}$ after 23 steps for this example. Furthermore we note that the matrix B possesses the maximal eigenvalue $\frac{1}{2}$. This is clear since

$$B = \begin{bmatrix} 0 & \frac{1}{4} & \frac{1}{4} & 0 \\ \frac{1}{4} & 0 & 0 & \frac{1}{4} \\ \frac{1}{4} & 0 & 0 & \frac{1}{4} \\ 0 & \frac{1}{4} & \frac{1}{4} & 0 \end{bmatrix}$$

has eigenvalues $\lambda_1 = \frac{1}{2}$, $\lambda_2 = -\frac{1}{2}$, $\lambda_3 = \lambda_4 = 0$. Consequently we have the spectral radius $\rho(B) = \frac{1}{2}$.

6.2 The Jacobi method

We shall now investigate the convergence of the Jacobi method as a special case of the following more general iteration:

$$x^{(r+1)} := Gx^{(r)} + v, \quad r = 0, 1, 2, \ldots \quad (6.48)$$

G is a (n, n) matrix and $v \in \mathbb{R}^n$, where

$$v := (I - G)A^{-1}b \quad (6.49)$$

and A is nonsingular.

Definition 6.50
The algorithm (6.48) is said to *converge to* $\bar{x}$ if the sequence $x^{(0)}, x^{(1)}, \ldots$, generated by (6.48), converges to $\bar{x}$ for all $x^{(0)}$.

Lemma 6.51
For a given matrix G we have

$$\lim_{r \to \infty} G^r w = 0, \quad \text{for all } w \in \mathbb{R}^n \quad (6.52)$$

if, and only if, $\rho(G) < 1$.

Proof:
Recall the following Theorem:
   Let $A \in \mathbb{C}^{n \times n}$ with spectral radius $\rho(A)$. Then for every $\epsilon > 0$ there exists a norm $\|\cdot\|$ such that $\|A\| \leq \rho(A) + \epsilon$.
   If $\rho(G) < 1$ then, according to this Theorem, there exists a matrix norm $\|\cdot\|_\beta$ with $\|G\|_\beta < 1$. For the vector norm $\|\cdot\|_\alpha$ corresponding to β it can be shown that

$$\|G^r w\|_\alpha \leq \|G^r\|_\beta \|w\|_\alpha \leq \|G\|_\beta^r \|w\|_\alpha . \quad (6.53)$$

As $r \to \infty$ the right-hand side of (6.53) converges to 0. Consequently (6.52) holds.
   On the other hand if (6.52) holds for all w then, obviously,

$$\lim_{r \to \infty} G^r = 0. \quad (6.54)$$

Since $A^m \to 0$ for $m \to \infty$ if, and only if, $\rho(A) < 1$, this together with (6.54) implies that $\rho(G) < 1$. Thus the Lemma is proved. $\square$

Theorem 6.55
If the algorithm (6.48) converges, then $\rho(G) < 1$, and for all $x^{(0)}$ the sequence $x^{(0)}, x^{(1)}, x^{(2)}, \ldots$, generated by (6.48), converges to the unique solution $\bar{x}$ of $Ax = b$.

Proof:
If the algorithm converges, then for $r \to \infty$ on both sides of (6.48), the equation

$$\bar{x} = G\bar{x} + v \qquad (6.56)$$

is obtained since

$$x^{(r)} \to \bar{x} \Rightarrow Gx^{(r)} \to G\bar{x}.$$

Here $\bar{x}$ is a limit vector to which the sequence generated by (6.48) converges for all $x^{(0)}$. Obviously

$$x^{(r+1)} - \bar{x} = G(x^{(r)} - \bar{x}) \qquad (6.57)$$

and

$$x^{(r)} - \bar{x} = G^r(x^{(0)} - \bar{x}). \qquad (6.58)$$

If the algorithm converges then by (6.58) and Lemma 6.51 the inequality

$$\rho(G) < 1$$

follows. In this case $v = (I - G)A^{-1}b$ and $\bar{x} = G\bar{x} + v$ imply $\bar{x} = \hat{x}$, where $\hat{x}$ is the exact solution.

If, on the other hand, $\rho(G) < 1$ then $I - G$ is nonsingular and the equation

$$x = Gx + v \qquad (6.59)$$

possesses a unique solution $\hat{x}$.

From (6.58) and Lemma 6.51 we deduce that $x^{(r)} \to \bar{x}$ for every $x^{(0)}$. The equation $v = (I - G)A^{-1}b$ implies that $\bar{x} = \hat{x}$. This proves the Theorem. □

It can now be shown that the Jacobi method converges if A is irreducible and weakly diagonally dominant. If μ is an eigenvalue of B then

$$det(B - \mu I) = det(D^{-1}C - \mu I) = 0. \qquad (6.60)$$

This implies that

$$det(\frac{1}{\mu}DD^{-1}C - \frac{1}{\mu}\mu DI) = det(D - \frac{1}{\mu}C) = 0. \qquad (6.61)$$

If A is weakly diagonally dominant and $|\mu| \geq 1$ then also $D - \frac{1}{\mu}C$ is weakly diagonally dominant. Moreover, if A is irreducible so also $D - \mu^{-1}C$ is irreducible. Then from Theorem 6.14 it follows that $det(D - \mu^{-1}C) \neq 0$.

This contradiction shows that μ can not be an eigenvalue of B for $|\mu| \geq 1$. Therefore $\rho(B) < 1$ and the Jacobi method converges.

Note 6.62
If A is positive definite the Jacobi method need not necessarily converge [62]. If, in addition, A is an L-matrix then $\rho(B) < 1$.

6.3 The Gauss-Seidel method

The Jacobi method frequently converges very slowly. A modification is the well-known *Gauss-Seidel method* which generally converges significantly faster.

In order to determine $x_i^{(r+1)}$ the Gauss-Seidel method requires

$$x_{i-1}^{(r+1)}, x_{i-2}^{(r+1)}, \ldots, x_1^{(r+1)}$$

instead of

$$x_{i-1}^{(r)}, x_{i-2}^{(r)}, \ldots, x_1^{(r)}.$$

Thus the iteration rule is

$$x_i^{(r+1)} := \sum_{k=1}^{i-1} b_{ik} x_k^{(r+1)} + \sum_{k=i+1}^{n} b_{ik} x_k^{(r)} + c_i. \tag{6.63}$$

For $n = 3$ this gives

$$\begin{aligned}
x_1^{(r+1)} &:= & & b_{12} x_2^{(r)} & + & b_{13} x_3^{(r)} & + & c_1 \\
x_2^{(r+1)} &:= & b_{21} x_1^{(r+1)} & & + & b_{23} x_3^{(r)} & + & c_2 \\
x_3^{(r+1)} &:= & b_{31} x_1^{(r+1)} & + b_{32} x_2^{(r+1)} & & & + & c_3.
\end{aligned} \tag{6.64}$$

Example 6.65
Consider the system of Example 6.47, namely:

$$\begin{bmatrix} x_1 \\ x_2 \\ x_3 \\ x_4 \end{bmatrix} = \begin{bmatrix} 0 & \frac{1}{4} & \frac{1}{4} & 0 \\ \frac{1}{4} & 0 & 0 & \frac{1}{4} \\ \frac{1}{4} & 0 & 0 & \frac{1}{4} \\ 0 & \frac{1}{4} & \frac{1}{4} & 0 \end{bmatrix} \cdot \begin{bmatrix} x_1 \\ x_2 \\ x_3 \\ x_4 \end{bmatrix} + \begin{bmatrix} 0 \\ 0 \\ 250 \\ 250 \end{bmatrix},$$

and apply the Gauss-Seidel method:

$$\begin{aligned}
x_1^{(r+1)} &:= \tfrac{1}{4} x_2^{(r)} + \tfrac{1}{4} x_3^{(r)} \\
x_2^{(r+1)} &:= \tfrac{1}{4} x_1^{(r+1)} + \tfrac{1}{4} x_4^{(r)} \\
x_3^{(r+1)} &:= \tfrac{1}{4} x_1^{(r+1)} + \tfrac{1}{4} x_4^{(r)} + 250 \\
x_4^{(r+1)} &:= \tfrac{1}{4} x_2^{(r+1)} + \tfrac{1}{4} x_3^{(r+1)} + 250.
\end{aligned}$$

Starting with the initial values

$$x_1^{(0)} = x_2^{(0)} = x_3^{(0)} = x_4^{(0)} = 0$$

we obtain an error $e_i^{(13)} \leq 10^{-4}$ after 13 iteration steps. This shows that the Gauss-Seidel method converges much more rapidly than the Jacobi method. At every iteration step r the errors

$$e_i^{(r)} := x_i^{(r)} - x_i, \quad r = 0, 1, 2, \ldots,$$

are reduced by a factor $\frac{1}{4}$; the corresponding factor for the Jacobi method is only $\frac{1}{2}$.

To investigate the convergence (6.63) is expressed in the form

$$x^{(r+1)} := Lx^{(r+1)} + Ux^{(r)} + c, \qquad (6.66)$$

where L is a strictly lower and U is a strictly upper triangular matrix with

$$L + U = B. \qquad (6.67)$$

For $n = 3$ we have

$$L := \begin{bmatrix} 0 & 0 & 0 \\ b_{21} & 0 & 0 \\ b_{31} & b_{32} & 0 \end{bmatrix}, \quad U := \begin{bmatrix} 0 & b_{12} & b_{13} \\ 0 & 0 & b_{23} \\ 0 & 0 & 0 \end{bmatrix}. \qquad (6.68)$$

$I - L$ is a lower triangular matrix with unit elements in its main diagonal. Therefore $det(I - L) = 1$ and thus $I - L$ is nonsingular. So $x^{(r+1)}$ in equation (6.66) is determinable:

$$x^{(r+1)} := Tx^{(r)} + (I - L)^{-1}c, \qquad (6.69)$$

where

$$T := (I - L)^{-1}U. \qquad (6.70)$$

From (6.36) the exact solution $\bar{x}$ of the original system $Ax = b$ will satisfy the equation

$$\bar{x} = B\bar{x} + c = L\bar{x} + U\bar{x} + c, \qquad (6.71)$$

and we have

$$\bar{x} = (I - L)^{-1}U\bar{x} + (I - L)^{-1}c = T\bar{x} + (I - L)^{-1}c. \qquad (6.72)$$

If $e^{(n)} := x^{(n)} - \bar{x}$ then

$$e^{(n+1)} = Te^{(n)}, \quad e^{(n)} = T^n e^{(0)}, \qquad (6.73)$$

i.e. the convergence of the Gauss-Seidel method depends on the eigenvalues of T just as the convergence of the Jacobi method depends on the eigenvalues of B.

Theorem 6.74
If A is irreducible and weakly diagonally dominant then $\rho(T) < 1$.

Proof:
If λ is an eigenvalue of T, then

$$\begin{aligned} det(T - \lambda I) &= det[(I - L)^{-1}U - \lambda I] \\ &= det[(I - L)^{-1}(U - \lambda(I - L))] \\ &= det[(I - L)^{-1}] \cdot det[U - \lambda(I - L)] \\ &= det(\lambda L + U - \lambda I) \\ &= 0. \end{aligned} \qquad (6.75)$$

6.3 The Gauss-Seidel method

If $|\lambda| \geq 1$ then $\lambda \neq 0$ and

$$det(I - \frac{1}{\lambda}U - L) = 0. \qquad (6.76)$$

Obviously the matrix $I - \frac{1}{\lambda}U - L$ is weakly diagonally dominant since the sum of the absolute values of the off-diagonal elements in the ith row does not exceed the value $\sum_{k=1}^{n} |b_{ik}|$, $k \neq i$.

Since A is irreducible, the matrix $V := I - \frac{1}{\lambda}U - L$ is irreducible and according to Theorem 6.14 it follows that V is nonsingular, and thus $det(T - \lambda I) \neq 0$.

Therefore λ is not an eigenvalue of T. This contradiction shows that $|\lambda| < 1$, if λ is an eigenvalue of T. $\square$

A further sufficient condition for the convergence of the Gauss-Seidel method is stated by the following Theorem:

Theorem 6.77
If A is a real positive definite matrix then $\rho(T) < 1$.

Proof:
The proof can be found in [54]. $\square$

An alternative proof for the convergence of the Gauss-Seidel method with a real positive definite matrix A is the following:

If P is a nonsingular matrix then the function

$$\|PAP^{-1}\|_2 \qquad (6.78)$$

possesses the properties of a matrix norm. Let

$$\|A\|_p := \|PAP^{-1}\|_2. \qquad (6.79)$$

The vector norm can be defined by analogy:

$$\|v\|_p := \|Pv\|_2. \qquad (6.80)$$

Finally

$$\|A\|_p := \sup_{v \neq 0} \frac{\|Av\|_p}{\|v\|_p}. \qquad (6.81)$$

The $A^{\frac{1}{2}}$-norm of T is considered:

$$\|T\|_{A^{\frac{1}{2}}} := \|A^{\frac{1}{2}}\|_2. \qquad (6.82)$$

$A^{\frac{1}{2}}$ is a unique positive definite matrix whose square is A. We have

$$\begin{aligned} T &= I - (I-L)^{-1}(I-L-U) \\ &= I - (I-L)^{-1}D^{-1}A \\ &= I - (D-C_L)^{-1}A \end{aligned} \qquad (6.83)$$

where $A := D - C_L - C_U$ and $D := diag(A)$. C_L is a strictly lower and C_U is a strictly upper triangular matrix. Thus

$$\begin{aligned} \bar{T} &= A^{\frac{1}{2}}TA^{-\frac{1}{2}} = I - A^{\frac{1}{2}}(D - C_L)^{-1}A^{\frac{1}{2}} \\ \bar{T}^T &= I - A^{\frac{1}{2}}(D - C_U)^{-1}A^{\frac{1}{2}}. \end{aligned} \quad (6.84)$$

Consequently one obtains

$$\begin{aligned} \bar{T}(\bar{T})^T &= I - A^{\frac{1}{2}}(D - C_L)^{-1}(D - C_U + D - C_L - A)(D - C_U)^{-1}A^{\frac{1}{2}} \\ &= I - A^{\frac{1}{2}}(D - C_L)^{-1}D(D - C_U)^{-1}A^{\frac{1}{2}} \\ &= I - [A^{\frac{1}{2}}(D - C_L)^{-1}D^{\frac{1}{2}}][A^{\frac{1}{2}}(D - C_L)^{-1}D^{\frac{1}{2}}]^T. \end{aligned} \quad (6.85)$$

Since $I - \bar{T}(\bar{T})^T$ is the product of a nonsingular matrix and its transpose, it follows that $I - \bar{T}(\bar{T})^T$ is positive definite and all eigenvalues of $\bar{T}(\bar{T})^T$ are less than 1. Since all eigenvalues are non-negative

$$\rho(T(T)^T) = \|T\|^2_{A^{\frac{1}{2}}} < 1 \quad (6.86)$$

and thus $\|T\|_{A^{\frac{1}{2}}} < 1$. Since the spectral radius is bounded by every matrix norm,

$$\rho(T) \le \|T\|_{A^{\frac{1}{2}}} < 1. \quad (6.87)$$

Definition 6.88
An $(n \times n)$ matrix $A := [a_{ik}]$ is said to be an *L-matrix* if $a_{ii} > 0$, $i = 1, 2, \ldots, n$, and $a_{ik} \le 0$ for $i, k = 1, 2, \ldots, n$ and $i \neq k$.

Definition 6.89
An $(n \times n)$ matrix $A := [a_{ik}]$ is said to be an *M-matrix* if A is a nonsingular L-matrix and all elements of A^{-1} are non-negative.

Note 6.90
Let $A := [a_{ik}]$, $i, k = 1, 2, \ldots, n$, be an L-matrix. Then the Gauss-Seidel method converges if, and only if, the Jacobi method converges. If both converge then the Gauss-Seidel method converges more rapidly than the Jacobi method in the sense that $\rho(T) \le \rho(B)$.

It can be shown that if A is an L-matrix, then both algorithms converge if, and only if, A is an M-matrix.

If L and U are given, the eigenvalues of T are obtainable by solving the equation

$$det(\lambda L + U - \lambda I) = 0. \quad (6.91)$$

For $n = 3$ we have

$$det \begin{bmatrix} -\lambda & b_{12} & b_{13} \\ \lambda b_{21} & -\lambda & b_{23} \\ \lambda b_{31} & \lambda b_{32} & -\lambda \end{bmatrix} = 0. \quad (6.92)$$

Example 6.93
Considering Example 6.47 we obtain:

$$\det \begin{bmatrix} -\lambda & \frac{1}{4} & \frac{1}{4} & 0 \\ \frac{1}{4}\lambda & -\lambda & 0 & \frac{1}{4} \\ \frac{1}{4}\lambda & 0 & -\lambda & \frac{1}{4} \\ 0 & \frac{1}{4}\lambda & \frac{1}{4}\lambda & -\lambda \end{bmatrix} = \lambda^3(\lambda - \tfrac{1}{4}) = 0.$$

Hence the eigenvalues of T are :

$$\lambda_1 = \frac{1}{4}, \lambda_2 = \lambda_3 = \lambda_4 = 0;$$

thus $\rho(T) = \frac{1}{4} = \rho(B)^2$.

The rate of convergence of the Gauss-Seidel method is twice that of the Jacobi method.

6.4 The SOR method

The Gauss-Seidel method can be modified so that the convergence is improved. The modification leads to the so-called *successive overrelaxation methods*. These are called *SOR methods* for short.

The SOR algorithm for solving the equation $Ax = b$ is defined by

$$x_i^{(r+1)} := \omega[\sum_{k=1}^{i-1} b_{ik} x_k^{(r+1)} + \sum_{k=i+1}^{n} b_{ik} x_k^{(r)} + c_i] + (1-\omega) x_i^{(r)}, \qquad (6.94)$$

where $i = 1, 2, \ldots, n$ and $r = 0, 1, 2, \ldots$. The real parameter ω is known as the *relaxation factor*. ω should be chosen so that the convergence is as rapid as possible. For $\omega = 1$, (6.94) is referred to as the Gauss-Seidel method.

For $n = 3$ one obtains

$$\begin{aligned} x_1^{(r+1)} &:= \omega(b_{12} x_2^{(r)} + b_{13} x_3^{(r)} + c_1) + (1-\omega) x_1^{(r)} \\ x_2^{(r+1)} &:= \omega(b_{21} x_1^{(r+1)} + b_{23} x_3^{(r)} + c_2) + (1-\omega) x_2^{(r)} \qquad (6.95) \\ x_3^{(r+1)} &:= \omega(b_{31} x_1^{(r+1)} + b_{32} x_2^{(r+1)} + c_3) + (1-\omega) x_3^{(r)}. \end{aligned}$$

In practice, the *Gauss-Seidel value* $\bar{x}_i^{(r+1)}$ is calculated first and then

$$x_i^{(r+1)} := x_i^{(r)} + \omega[\bar{x}_i^{(r+1)} - x_i^{(r)}]. \qquad (6.96)$$

This process is a kind of extrapolation or *overrelaxation*.

Example 6.97
Applying the SOR method to Example 6.47 one obtains for $r = 0, 1, 2, \ldots$:

$$\begin{aligned} x_1^{(r+1)} &:= \omega(\tfrac{1}{4} x_2^{(r)} + \tfrac{1}{4} x_3^{(r)}) & + (1-\omega) x_1^{(r)} \\ x_2^{(r+1)} &:= \omega(\tfrac{1}{4} x_1^{(r+1)} + \tfrac{1}{4} x_4^{(r)}) & + (1-\omega) x_2^{(r)} \\ x_3^{(r+1)} &:= \omega(\tfrac{1}{4} x_1^{(r+1)} + \tfrac{1}{4} x_4^{(r)} & + 250) + (1-\omega) x_3^{(r)} \\ x_4^{(r+1)} &:= \omega(\tfrac{1}{4} x_2^{(r+1)} + \tfrac{1}{4} x_3^{(r+1)} & + 250) + (1-\omega) x_4^{(r)}. \end{aligned} \qquad (6.98)$$

With the initial values

$$x_1^{(0)} = x_1^{(2)} = x_3^{(0)} = x_4^{(0)} = 0$$

and $\omega := 1.072$ we obtain an error $e_i^{(8)} \leq 10^{-4}$ after 8 iteration steps. Evidently the convergence is more rapid than that of the Gauss-Seidel method.

Generally, at every iteration step the error is reduced by a minimum factor $\frac{1}{10}$. On the other hand, the quotient $e_i^{(r+1)}/e_i^{(r)}$ is often more irregular than that of the Gauss-Seidel method since several eigenvalues of the matrix assigned to the SOR algorithm are often of the same multiplicity.

For the investigation of the convergence we look at the following form of the SOR algorithm:

$$x^{(r+1)} := \omega(Lx^{(r+1)} + Ux^{(r)} + c) + (1-\omega)x^{(r)}, \quad r = 0, 1, 2, \ldots . \tag{6.99}$$

Since $I - \omega L$ is nonsingular, one obtains for $x^{(r+1)}$:

$$x^{(r+1)} := T_\omega x^{(r)} + (I - \omega L)^{-1} \omega c, \tag{6.100}$$

where

$$T_\omega := (I - \omega L)^{-1}(\omega U + (1-\omega)I). \tag{6.101}$$

The exact solution $\bar{x}$ of the original system $Ax = b$ satisfies the equation

$$\bar{x} = B\bar{x} + c = L\bar{x} + U\bar{x} + c \tag{6.102}$$

and thus we have

$$\omega \bar{x} = \omega L \bar{x} + \omega U \bar{x} + \omega c \tag{6.103}$$

and

$$\bar{x} = \omega L \bar{x} + \omega U \bar{x} + \omega c + (1-\omega)\bar{x}. \tag{6.104}$$

This is equivalent to

$$\bar{x} = T_\omega \bar{x} + (I - \omega L)^{-1} \omega c. \tag{6.105}$$

Hence

$$e^{(r+1)} = T_\omega e^{(r)}, \quad e^{(r)} = T_\omega^r e^{(0)}; \tag{6.106}$$

i.e. the convergence of the SOR algorithm depends on the eigenvalues of T_ω.

Note 6.107
If A is irreducible, weakly diagonally dominant and $0 < \omega < 1$, then $\rho(T_\omega) < 1$ (see [62]). This result is of small practical importance since more a rapid convergence is usually achieved for $\omega > 1$.

How to test the irreducibility of a matrix will be discussed in Chapter 7 where graphs and matrices are treated. It was shown in [26] that if the SOR algorithm is to converge, then ω cannot lie outside of the interval $(0,2)$.

6.4 The SOR method

Theorem 6.108
If $\rho(T_\omega) < 1$ then $0 < \omega < 2$.

Proof:
The proof can be found in [62]. □

The following results are summarized:

- If A is real and positive definite and $0 < \omega < 2$, then the SOR method will converge (according to [36]).

- If the diagonal elements of a symmetric matrix A are positive, and the SOR method converges, then A is positive definite and $0 < \omega < 2$.

If L and U are given, the eigenvalues of T_ω may be calculated from

$$\det(\lambda L + U - \frac{\lambda + \omega - 1}{\omega} I) = 0, \quad \text{where } \omega \neq 0. \tag{6.109}$$

Example 6.110
For the system of Example 6.47 one obtains

$$\det \begin{bmatrix} -a & \frac{1}{4} & \frac{1}{4} & 0 \\ \frac{1}{4}\lambda & -a & 0 & \frac{1}{4} \\ \frac{1}{4}\lambda & 0 & -a & \frac{1}{4} \\ 0 & \frac{1}{4}\lambda & \frac{1}{4}\lambda & -a \end{bmatrix} = 0, \tag{6.111}$$

where $a := \frac{1}{\omega}(\lambda + \omega - 1)$. (6.111) can be solved by

$$\det \begin{bmatrix} -a & \frac{1}{4}\sqrt{\lambda} & \frac{1}{4}\sqrt{\lambda} & 0 \\ \frac{1}{4}\sqrt{\lambda} & -a & 0 & \frac{1}{4}\sqrt{\lambda} \\ \frac{1}{4}\sqrt{\lambda} & 0 & -a & \frac{1}{4}\sqrt{\lambda} \\ 0 & \frac{1}{4}\sqrt{\lambda} & \frac{1}{4}\sqrt{\lambda} & -a \end{bmatrix} = a^2(a^2 - \frac{1}{4}\lambda) \tag{6.112}$$

so that $(\lambda + \omega - 1)^2 = 0$ or

$$(\lambda + \omega - 1)^2 = \frac{1}{4}\omega^2 \lambda. \tag{6.113}$$

For $\omega = 1.1$ the following eigenvalues of $T_{1.1}$ are obtained:

$$-0.1, \quad -0.1, \quad 0.05125 + 0.08587i, \quad 0.05125 - 0.08587i$$

Evidently

$$\rho(T_{1.1}) = 0.1 < \rho(T) = 0.25. \tag{6.114}$$

Similary, for $\omega := 1.072$ we have the following eigenvalues

$$-0.072, \quad -0.072, \quad 0.07165 + 0.00711i, \quad 0.07165 - 0.00711i$$

and

$$\rho(T_{1.072}) = 0.072.$$

It can be shown that $\omega := 1.072$ minimizes the function $\rho(T_\omega)$.

In more recent work [11] the problem of determining an optimal ω is intensively studied.

The iteration algorithm quoted so far may be used for block matrices. Thus, in the decomposed form of A, the matrix D is of the form

$$D := \begin{bmatrix} D_1 & & 0 \\ & \ddots & \\ 0 & & D_m \end{bmatrix}.$$

The matrices D_i, $i = 1, \ldots, m$ are appropriately chosen square submatrices. Newer block structures are investigated in [12]:

$$D := \begin{bmatrix} \beta_1 & \gamma_1 & 0 & \cdots & & 0 & \alpha_1 \\ \alpha_1 & \beta_2 & \gamma_2 & \ddots & & & 0 \\ 0 & \alpha_3 & \beta_3 & \gamma_3 & & & \vdots \\ \vdots & & \ddots & \ddots & \ddots & & \vdots \\ \vdots & & & & \ddots & & 0 \\ 0 & & & & \ddots & \ddots & \gamma_{n-1} \\ \gamma_n & 0 & \cdots & & 0 & \alpha_n & \beta_n \end{bmatrix}.$$

These are encountered when dealing with self-adjoint, elliptic partial differential equations.

6.5 The SOR method with block matrices

First we consider the system of equations

$$Ax = b, \quad A \in \mathbb{R}^{n \times n}, \quad x, b \in \mathbb{R}^n, \tag{6.115}$$

where A is completely filled and of the form

$$\begin{bmatrix} A_{11} & A_{12} & \cdots & A_{1m} \\ A_{21} & A_{22} & \cdots & A_{2m} \\ \vdots & \vdots & & \vdots \\ A_{m1} & A_{m2} & \cdots & A_{mm} \end{bmatrix} \cdot \begin{bmatrix} x^1 \\ x^2 \\ \vdots \\ x^m \end{bmatrix} = \begin{bmatrix} b^1 \\ b^2 \\ \vdots \\ b^m \end{bmatrix} \tag{6.116}$$

with square submatrices

$$A_{ii} \in \mathbb{R}^{r_i \times r_i}, \quad x^i, b^i \in \mathbb{R}^{r_i}, \quad i = 1, 2, \ldots, m.$$

Then for k iterations the following block iterations in the SOR method without additive splitting are obtained:

With $A = I - L - U$, i.e. $D = I$ the equation

$$(I - \omega L)x^{(k+1)} = \omega b + [(1-\omega)I + \omega U]x^{(k)}$$

will follow or, in component notation, we have

$$\begin{aligned} x_i^{(k+1)} = &\; \omega(b_i - a_{i1}x_1^{(k+1)} - \ldots - a_{i,i-1}x_{i-1}^{(k+1)}) + (1-\omega)x_i^{(k)} \\ &+ \omega(-a_{i,i+1}x_{i+1}^{(k)} - \ldots - a_{in}x_n^{(k)}). \end{aligned} \tag{6.117}$$

6.5 The SOR method with block matrices

Hence

$$A_{ii}x^i_{(k+1)} = \omega(b^i - A_{i1}x^1_{(k+1)} - \ldots - A_{i,i-1}x^{i-1}_{(k+1)})$$
$$+ (1-\omega)A_{ii}x^i_{(k)} + \omega(-A_{i,i+1}x^{i+1}_{(k)} - \ldots - A_{im}x^m_{(k)}), \quad (6.118)$$
$$i = 1, \ldots, m.$$

The right-hand sides of (6.118) are calculated and then the systems of equations for the approximation solutions $x^i_{(k+1)}$ of the partial solutions x^i are solved.

In order to carry out the iteration process the matrices A_{ii} are decomposed into the factors L_{ii} and U_{ii} so that

$$A_{ii} = L_{ii}U_{ii}. \quad (6.119)$$

The systems are solved by noting the sparseness of the matrices.

A variant of (6.118) is the following algorithm:

$$x^i_{(k+1)} := \omega(\hat{b}^i - \hat{A}_{i1}x^1_{(k+1)} - \ldots - \hat{A}_{i,i-1}x^{i-1}_{(k)})$$
$$+ (1-\omega)x^i_{(k)} + \omega(-\hat{A}_{i,i+1}x^{i+1}_{(k)} - \ldots - \hat{A}_{im}x^m_{(k)}), \quad (6.120)$$

where

$$\hat{b}^i = A_{ii}^{-1}b^i \text{ and } \hat{A}_{ij} = A_{ii}^{-1}A_{ij}$$

(see general SOR method).

The inverse is calculated indirectly by solving the systems

$$A_{ii}\hat{A}_{ij} = A_{ij}$$
$$A_{ii}\hat{b}^i = b^i,$$

where $\hat{A}_{ij}$ and $\hat{b}^i$ are the solutions sought. Then, by making use of the SOR block method, the system $\hat{A}x = \hat{b}$ is solved:

$$\begin{bmatrix} I & \hat{A}_{12} & \cdots & \hat{A}_{1m} \\ \hat{A}_{21} & I & \cdots & \hat{A}_{2m} \\ \vdots & & & \vdots \\ \hat{A}_{m1} & \hat{A}_{m2} & \cdots & I \end{bmatrix} \cdot \begin{bmatrix} x^1 \\ x^2 \\ \vdots \\ x^m \end{bmatrix} = \begin{bmatrix} \hat{b}^1 \\ \hat{b}^2 \\ \vdots \\ \hat{b}^m \end{bmatrix}. \quad (6.121)$$

If $A = A^T$ and A is positive definite, then also $A_{ii} = A_{ii}^T$ and A_{ii} is positive definite. In this special case the Cholesky decomposition is applied to the individual blocks A_{ii}:

$$A_{ii} = L_{ii}L_{ii}^T. \quad (6.122)$$

It can then be shown that an SOR block iteration is equivalent to an SOR iteration for the system

$$\hat{A}\hat{x} = \hat{b}, \quad (6.123)$$

where $\hat{A} = \hat{L}^{-1}A\hat{L}^{-T}$, $\hat{x} = \hat{L}^Tx$, $\hat{b} = \hat{L}^{-1}b$ and

$$\hat{L} = \begin{bmatrix} L_{11} & & & \\ & L_{22} & & 0 \\ & & \ddots & \\ 0 & & & L_{mm} \end{bmatrix}.$$

Since $A = A^T$ and A is positive definite, the SOR method is applied to symmetric positive definite subsystems. For $0 < \omega < 2$ the convergence of the method is ensured.

Note 6.124
The method can be reduced to the case of a tridiagonal block matrix A.

6.6 Convergence criteria for SOR methods

In more recent publications [25] additional convergence criteria are investigated which incorporate the former conditions and are valid for a further class of sparse matrices. Here sparseness is not a precondition. These matrices, dealt with more precisely in what follows, will be called *generalized diagonally dominant* matrices.

The well-known sufficient conditions derived from the coefficient matrix A which ensure the convergence of the Gauss-Seidel method are positive definiteness and diagonal dominance. Now relations between the row and the column scaling of the coefficient matrix, the diagonal dominance and the convergence of the above iterative algorithms can be established.

The above-mentioned L-matrices (see Definition 6.88) which arise frequently in applied mathematics are regarded and treated as a special class of matrices. Convergent irreducible L-matrices may be characterized by strict diagonal dominance under row or column scaling.

Supplementary to the Definition 6.7 of weakly diagonally dominant matrices the following definitions are introduced:

Definition 6.125
A matrix $A := [a_{ik}]$, $A \in \mathbb{C}^{n \times n}$, is said to be *strictly diagonally dominant* if

$$|a_{ii}| > \sum_{\substack{k=1 \\ k \neq i}}^{n} |a_{ik}| \quad \text{for all } i.$$

Definition 6.126
Let $A := [a_{ik}]$, $A \in \mathbb{C}^{n \times n}$. If a scaling of the columns (rows) of A with scalars $\alpha \neq 0$ exists in such a way that the transformed matrix is strictly diagonally dominant in the rows (columns), then A is said to possess *generalized diagonal dominance* in the rows (columns).

Definition 6.127
Let $A := [a_{ik}]$ and $B := [b_{ik}]$ be two $(n \times r)$ matrices. Then $A \geq B$ if $a_{ik} \geq b_{ik}$ for all $1 \leq i \leq n$, $1 \leq k \leq r$. If 0 is the zero matrix and $A \geq 0$, then A is a *non-negative matrix*. By analogy, a *positive matrix* $A > 0$ is defined.

Note 6.128
When $A := [a_{ik}]$, $a_{ik} \in \mathbb{C}$, then $|A|$ denotes the non-negative matrix with the elements $|a_{ik}|$.

6.6 Convergence criteria for SOR methods

Clearly, the rows of the coefficient matrix may be scaled by arbitrary scalars $\alpha \neq 0$ without affecting the convergence properties. Therefore

$$A := I + L + U \tag{6.129}$$

is scaled in such a way that the main diagonal elements are 1. Here L and U are strictly lower or strictly upper triangular matrices, respectively.

First a Lemma is given for the meaningful description of matrix scaling and diagonal dominance.

Lemma 6.130
Let $A := I + L + U$ be an $(n \times n)$ matrix. Then A possesses generalized diagonal dominance in the rows or columns respectively if, and only if, there exist positive vectors v and w satisfying

$$(I - |L| - |U|)v = w \tag{6.131}$$

or

$$(I - |L| - |U|)^T v = w, \tag{6.132}$$

respectively.

Proof:
Let v be a vector with $v_i \neq 0$ for $i = 1, 2, \ldots, n$. v_i is assumed to scale the ith column of A. The rows of the scaled matrix are strictly diagonally dominant if, and only if,

$$|v_i| - \sum_{\substack{k=1 \\ k \neq i}}^{n} |v_k| \cdot |a_{ik}| > 0, \quad \text{for } i = 1, 2, \ldots, n. \tag{6.133}$$

But this is exactly the condition for the existence of 2 positive vectors v, w satisfying equation (6.131). By analogy the result for row scaling will follow. □

For the development of convergence conditions the following Lemma is required:

Lemma 6.134
Let $A \geq 0$ be an irreducible $(n \times n)$ matrix. Then A possesses a real positive eigenvalue equal to the spectral radius $\rho(A)$ and a corresponding real positive eigenvector.

Proof:
This result follows from the Perron-Frobenius theory [62]. □

With these preliminary considerations the following Theorem can be formulated:

Theorem 6.135
Let $A := I - B$ be an irreducible L-matrix. Then the Gauss-Seidel method and the Jacobi method will converge for A if A satisfies the condition of the generalized diagonal dominance in the rows.

Proof:
The well-known conclusion of the Stein-Rosenberg theory [53] applied to the matrix A states that the Gauss-Seidel method converges if, and only if, the Jacobi algorithm converges. By supposition the Jacobi iteration matrix B is irreducible and $B \geq 0$. Lemma 6.134 asserts that B possesses a real positive eigenvalue λ equal to $\rho(B)$ and a related positive eigenvector z. Then

$$Bz = \lambda z \qquad (6.136)$$

implies

$$Az = Iz - \lambda z = (1 - \lambda)z, \qquad (6.137)$$

where $0 < \lambda < 1$ if the Jacobi method converges. Since by assumption $A = I - B$ where $B \geq 0$, there exist positive vectors z, v satisfying the equation

$$(I - |B|)z = v.$$

Then Lemma 6.130 shows that A satisfies the condition of generalized diagonal dominance in the rows. □

Note 6.138
A similar sufficient condition for the convergence is the generalized diagonal dominance in the columns.

This result may be carried over to SOR methods.

Theorem 6.139
(i) Let $A := I - L - U$ be an irreducible $(n \times n)$ matrix, with $L \geq 0$ and $U \geq 0$. If A satisfies the condition of generalized diagonal dominance in rows, then the SOR method converges for $0 < \omega \leq 1$ for A.
(ii) Let $\bar{A} = I - L + U$ be an irreducible $(n \times n)$ matrix, with $L \geq 0$ and $U \geq 0$. If A satisfies the condition of generalized diagonal dominance in rows, then the SOR method with $\omega > 1$ converges for $\bar{A}$.

Proof:
According to (6.101) the SOR algorithm for A possesses the following iteration matrix T_ω:

$$\begin{aligned} T_\omega &:= (I - \omega L)^{-1}(\omega U + (1 - \omega)I) \\ &= (I + \omega L + \omega^2 L^2 + \ldots)(\omega U + (1 - \omega)I). \end{aligned}$$

Since $L \geq 0$ and $U \geq 0$, T_ω is also a non-negative matrix for $0 < \omega \leq 1$.

6.6 Convergence criteria for SOR methods

Since $I - L - U$ is irreducible for $\omega \neq 1$, so T_ω is also irreducible. The case $\omega = 1$ is included in Theorem 6.135.

According to Lemma 6.134 T_ω possesses a real positive eigenvalue λ equal to the spectral radius $\rho(T_\omega)$ with corresponding positive eigenvector z. By the definition of T_ω the equation $T_\omega z = \lambda z$ yields

$$[(1-\omega)I + \omega U]z = (I - \omega L)\lambda z. \tag{6.140}$$

Since $L \geq 0$ and $U \geq 0$ it follows that

$$\omega(I - |L| - |U|)z = (1-\lambda)(I - \omega L)z. \tag{6.141}$$

Here $0 < \lambda < 1$ if the iteration converges, and z is a positive vector. The right-hand side of (6.141) is a positive vector because $\lambda > 0$, and (6.140) shows that $(I - \omega L)z$ is positive. According to Lemma 6.130 in conjunction with (6.141) the matrix A satisfies the condition of generalized diagonal dominance in the rows. Part (ii) of the Theorem is proved analogously. □

Note 6.142
By row or column scaling, irreducible, weakly diagonally dominant matrices that arise when solving partial differential equations may be transformed into strictly diagonally dominant matrices. Thus they trivially fulfil the sufficient conditions of Theorem 6.135 and Theorem 6.139.

By scaling in rows and columns one arrives at sufficient conditions for the convergence of SOR methods. First, the following notations are given:
Let

$$A := D + L + U \tag{6.143}$$

be an $(n \times n)$ matrix, with $D := diag(A) = [a_{ii}]$, $a_{ii} \neq 0$ for $i = 1, 2, \ldots, n$. Let L be a strictly lower, and U a strictly upper, triangular matrix.

Theorem 6.144
Let (6.143) apply to $A := [a_{ik}]$, $i, k = 1, 2, \ldots, n$. Let $\bar{A} := [\bar{a}_{ik}]$ be that matrix which results from scaling the rows and columns of A by arbitrary scalars different from zero. Then the SOR iteration matrices for A and $\bar{A}$ possess the same eigenvalues.

Proof:
The SOR iteration matrix for A is

$$T_\omega := (D + \omega L)^{-1}[(1-\omega)D - \omega U]. \tag{6.145}$$

Let P and Q be diagonal matrices with nonzero diagonal elements. Then by construction

$$\bar{A} := PAQ = PDQ + PLQ + PUQ$$

and consequently the SOR iteration matrix for $\bar{A}$ is

$$\begin{aligned}\bar{T}_\omega &:= [P(D + \omega L)Q]^{-1}[(1-\omega)PDQ - \omega PUQ] \\ &= Q^{-1}T_\omega Q,\end{aligned} \tag{6.146}$$

i.e. T_ω and $\bar{T}_\omega$ possess the same eigenvalues. □

In this context the following lemmas are mentioned:

Lemma 6.147
The SOR algorithm converges for A if, and only if, it converges for $\bar{A}$. In both cases the *rate of convergence* is the same.

Lemma 6.148
If the transformed matrix $\bar{A}$ is diagonally dominant, then the SOR algorithm converges for $0 < \omega < \bar{\omega}$ with

$$\bar{\omega} := \frac{2}{1 + \min(s_1, s_2)}, \tag{6.149}$$

and

$$s_1 := \max_i \left(\frac{\sum_{k \neq i} |\bar{a}_{ik}|}{|\bar{a}_{ii}|} \right), \quad s_2 := \max_k \left(\frac{\sum_{i \neq k} |\bar{a}_{ik}|}{|\bar{a}_{ii}|} \right).$$

This also implies the convergence of the SOR algorithm for A with $0 < \omega < \bar{\omega}$.

Example 6.150

$$|A| := \begin{bmatrix} 2.0 & 4.0 & 0.6 \\ 0.2 & 2.0 & 0.2 \\ 0.6 & 0.8 & 1.0 \end{bmatrix}.$$

Scaling of the 1st row by $\alpha := \frac{1}{2}$, the 2nd row by $\alpha := 2$ and the 2nd column by $\beta := \frac{1}{4}$ yields

$$\bar{A} := \begin{bmatrix} 1.0 & 0.5 & 0.3 \\ 0.4 & 1.0 & 0.4 \\ 0.6 & 0.2 & 1.0 \end{bmatrix}.$$

Hence by (6.149) $\bar{\omega} := 1.1$, i.e. the SOR algorithm converges for $0 < \omega < 1.1$.

It is known from [36] that for a Hermitian matrix A with positive diagonal elements the positive definiteness of A constitutes a necessary and sufficient condition for the convergence of the SOR algorithm. To complement Theorem 6.135 the following Theorem holds:

Theorem 6.151
Let $M \geq 0$ and let $A := I - (M + M^T)$ be an irreducible $(n \times n)$ matrix. If A is positive definite, then the condition of generalized diagonal dominance applies to A.

Proof:
The matrix $M + M^T$ is irreducible and $M + M^T \geq 0$. Then, according to Lemma 6.134, the matrix possesses a positive eigenvector z. This vector z is also an eigenvector of A; i.e.

$$Az = \lambda z. \tag{6.152}$$

Here λ and z are real and positive, i.e.

$$(I - |M| - |M^T|)z = v, \tag{6.153}$$

where z and v are positive vectors. According to Lemma 6.130 the condition of generalized diagonal dominance applies to A. □

6.6 Convergence criteria for SOR methods

Within the scope of these considerations we state another generalization of the property of a Hermitian matrix A with positive diagonal elements, i.e. A is positive definite if it is diagonally dominant. For this purpose the following definition and lemma are required [62].

Definition 6.154
A matrix A is said to be *N-stable* if all eigenvalues of A possess positive real parts.

Lemma 6.155
A real matrix H is N-stable if, and only if, there exists a positive definite matrix Q such that the matrix $K := HQ + QH^T$ is positive definite.

Thus we can formulate the following Theorem:

Theorem 6.156
Let $A := [a_{ik}]$, $A \in \mathbb{C}^{n \times n}$, with all real, positive diagonal elements. If the condition of generalized diagonal dominance is satisfied by A in rows or columns, then A is N-stable.

In practice the scaling and its influence on convergence must be carefully investigated. Moreover, the question of minimizing the number of operations in the scaling plays an interesting role in large sparse matrix problems.

7 Graphs and matrices

Graphs have twofold meaning in the context of our problems. Investigations of economic life reveal multiple connections between individuals, groups of persons and objects. Consider, for example, the influence that the employers of a company can exercise, directly or indirectly, on each other, in particular through exchange of information: or, another example, the dependence of the production of some article on the availability of another: or the sequence of operations in a building project and the importance of the availability of the right materials at the correct moment: or, again, the communication links that connect one town with another.

For the description and investigation of such situations the mathematical concepts of graph theory are appropriate. In the next section a representation of graphs by matrices will be given. On the other hand, certain matrices (adjacency matrices) can also be regarded as graphs so that the auxiliary material of graph theory may be utilized.

A graph G can be represented by a finite set $\mathcal{V} \neq \emptyset$ of p points and a set $\mathcal{X}$, containing q non-ordered pairs of elements of $\mathcal{V}$; thus $(u, v) \in \mathcal{X}$, where $u, v \in \mathcal{V}$. Each of these pairs (u, v) is connected by a straight line.

Example 7.1
Graphs with 4 points:

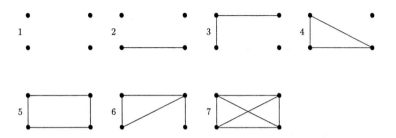

The points are called *vertices*, and the straight lines *edges*. (This definition is not unique!) When the straight lines have a direction they may be regarded as *arrows* and one speaks of a *directed graph* G:

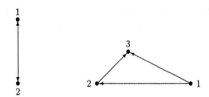

For example, if the flow of information within an enterprise is to be investigated the members of the firm may be represented by points in the plane. Then, when the person a can pass information directly to person b an arrow is drawn from point a to point b.

It is possible for 2 vertices to be connected by several edges (by several arrows of the same direction = *parallel arrows*) or for one vertex to be connected with itself (=*loop*).

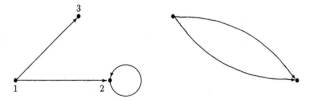

Many problems can be restricted to directed graphs having no loops and no parallel arrows. Graphs of that kind are called digraphs.

Definition 7.2
Let $\mathcal{X}$ and $\mathcal{V}$ be two sets and $f : \mathcal{X} \to \mathcal{V}$, $g : \mathcal{X} \to \mathcal{V}$ two mappings with the following properties:
$$\mathcal{V} \text{ is finite and not empty;} \qquad (7.3)$$
for all $x, y \in \mathcal{X}$ the implications
$$\left.\begin{array}{r}f(x) = f(y) \\ g(x) = g(y)\end{array}\right\} \Rightarrow x = y \qquad (7.4)$$
are true.
For all $x \in \mathcal{X}$
$$f(x) \neq g(x) \qquad (7.5)$$
holds.
Then the quadruplet $\mathsf{G} = \{\mathcal{X}, \mathcal{V}, f, g\}$ is called a *digraph*; the elements of $\mathcal{X}$ are called *arrows* and the elements of $\mathcal{V}$ are named *vertices*.

Let $x \in \mathcal{X}$ be an arrow; $f(x) \in \mathcal{V}$ is called the *initial vertex* of $\mathcal{X}$, and $g(x) \in \mathcal{V}$ is called the *final vertex*.

Note 7.6

- (7.4) states that two arrows having the same initial vertices and final vertices are the same, i.e. parallel arrows are excluded.

- (7.5) states that no arrows of a digraph possess an initial vertex and a final vertex which coincide, i.e. loops are excluded.

- Two graphs $\mathsf{G}_i = (\mathcal{X}_i, \mathcal{V}_i, f_i, g_i)$, $i = 1, 2$, are called *equal* if, and only if, $\mathcal{X}_1 = \mathcal{X}_2, \mathcal{V}_1 = \mathcal{V}_2, f_1 = f_2$ and $g_1 = g_2$.

Theorem 7.7
Let $\mathsf{G} = \{\mathcal{X}, \mathcal{V}, f, g\}$ be a digraph. Then there are finitely many arrows, i.e. $\mathcal{X}$ is finite.

Example 7.8
Let $\mathcal{X} := \{x_1, x_2, x_3, x_4\}$ and $\mathcal{V} := \{v_1, v_2, v_3, v_4, v_5\}$ be two sets and
$$f := \begin{cases} x_1 \to v_1 \\ x_2 \to v_1 \\ x_3 \to v_4 \\ x_4 \to v_5 \end{cases} \qquad g := \begin{cases} x_1 \to v_2 \\ x_2 \to v_5 \\ x_3 \to v_2 \\ x_4 \to v_1 \end{cases}$$
two mappings of $\mathcal{X}$ into $\mathcal{V}$.

$G = (\mathcal{X}, \mathcal{V}, f, g)$ is a digraph:

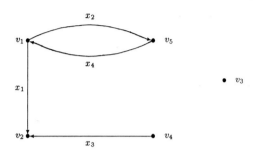

We now give some characterizations of special digraphs and the representation of digraphs by matrices.

Definition 7.9
A digraph $(\mathcal{X}, \mathcal{V}, f, g)$ is said to be *symmetric*, if for all $x \in \mathcal{X}$ there is an element $y \in \mathcal{X}$ such that
$$f(x) = g(y) \land g(x) = f(y)$$
holds, i.e. for each arrow there is an arrow having a reversed direction.

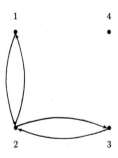

Definition 7.10
A digraph $(\mathcal{X}, \mathcal{V}, f, g)$ is called *asymmetrical*, if for all $x \in \mathcal{X}$ and all $y \in \mathcal{X}$
$$f(x) \neq g(y) \lor g(x) \neq f(y);$$
i.e. for no arrow is there an arrow with reversed direction.

Definition 7.11
A digraph is called *transitive*, if for all $x \in \mathcal{X}$ and all $y \in \mathcal{X}$ there exists an element $z \in \mathcal{X}$ such that

$$(g(x) = f(y) \wedge f(x) \neq g(y)) \Rightarrow (f(z) = f(x) \wedge g(z) = g(y));$$

i.e. for any two arrows which do not meet in the reverse sense there exists a *composition arrow*.

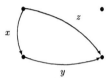

Definition 7.12
A digraph is called *complete*, if for all $u \in \mathcal{V}$ and all $v \in \mathcal{V}$ where $u \neq v$, there exists an element $x \in \mathcal{X}$ such that

$$(f(x) = u \wedge g(x) = v) \vee (f(x) = v \wedge g(x) = u);$$

i.e. each pair of vertices is connected by at least one arrow.

Definition 7.13
Let $D = (\mathcal{X}, \mathcal{V}, f, g)$ be a digraph whose vertices are enumerated by aid of numbers $1, 2, \ldots, n : \mathcal{V} = \{v_1, \ldots, v_n\}$. The adjacency matrix $A(D)$ of the digraph D relating to the given enumeration of its vertices is an $(n \times n)$ matrix $[a_{ik}]$ where

$$a_{ik} := \begin{cases} 1, & \text{if there is an arrow from } v_i \text{ to } v_k, \quad i, k = 1, 2, \ldots, n \\ 0 & \text{otherwise.} \end{cases}$$

Thus $a_{ik} = 1$ if, and only if, there is an arrow $x \in \mathcal{X}$ for which $f(x) = v_i$ and $g(x) = v_k$.

Example 7.14

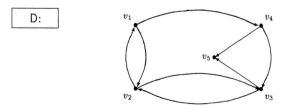

7 Graphs and matrices

A(D):

from \ to	v_1	v_2	v_3	v_4	v_5
v_1	0	1	0	1	0
v_2	1	0	1	0	0
v_3	0	1	0	0	1
v_4	0	0	1	0	1
v_5	0	0	0	0	0

If corresponding rows and columns of an adjacency matrix are interchanged simultaneously this indicates merely a renumbering of the vertices of the digraph. This invariant property under multiplication by permutation matrices, for example, is utilized when studying the fill-in of the Gaussian elimination process.

Our goal was to investigate sparse matrix problems by means of graph theory. For this purpose let $A := [a_{ik}]$, $i, k = 1, 2, \ldots, n$ be a matrix. We consider n points of the plane numbered $1, 2, \ldots, n$. For each $i, k \in \{1, \ldots, n\}$ an arrow is drawn from point i to point k, if $a_{ik} \neq 0$. In this way we obtain a digraph corresponding to the matrix A.

Now by graph theory we can obtain an irreducibility criterion for a matrix A. To do this we must first introduce some ideas concerning the connectedness of graphs.

We will show that the matrix A is irreducible if, and only if, the corresponding graph is connected in the following sense:
Let $i, k \in \{1, \ldots, n\}$; $i \neq k$.
Then there must exist indices $i_1, i_2, \ldots, i_s \in \{1, \ldots, n\}$ so that arrows from i_1 to i_2, i_2 to $i_3, \ldots, i_{s-1}$ to i_s can be drawn.

Example 7.15
The following matrix is not irreducible; the graph is not connected in the sense named above (there is no arrow from 3 to 1!).

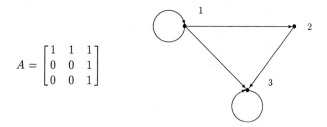

$$A = \begin{bmatrix} 1 & 1 & 1 \\ 0 & 0 & 1 \\ 0 & 0 & 1 \end{bmatrix}$$

Definition 7.16
Let $D = (\mathcal{X}, \mathcal{V}, f, g)$ be a digraph and $u, v \in \mathcal{V}$ two different vertices. If v can be reached from u, i.e. if there is a sequence of arrows $(x_1, x_2, \ldots, x_s)$ with $f(x_1) = u$ and $g(x_s) = v$, the *ordered pair* $(u, v) \in \mathcal{V} \times \mathcal{V}$ is said to be *strictly connected*.

Definition 7.17
A digraph is called *quasi-strictly connected*, if for two arbitrary different vertices $u, v \in \mathcal{V}$ the pair (u, v) or the pair (v, u) is strictly connected. A digraph is said to be *strictly connected*, if (u, v) and (v, u) are strictly connected for all $u, v \in \mathcal{V}$ where $u \neq v$.

The following criterion can now be formulated.

Theorem 7.18
A quadratic matrix A is *irreducible* if, and only if, its corresponding digraph is strictly connected.

Proof:
Using the Definition 7.13 of an adjacency matrix $A(\mathrm{D})$ and Theorem 6.12 of Chap. 6 the assertion follows. □

As an application of this theory we consider the eigenvalue problem
$$Ax = \lambda x, \tag{7.19}$$
where A is a sparse matrix. The task is to condense the matrix A and to solve the problem by block triangulation. A permutation matrix P is constructed so that
$$PAP^{-1} := \begin{bmatrix} A_{11} & & B \\ & \ddots & \\ 0 & & A_{rr} \end{bmatrix} \tag{7.20}$$
holds for arbitrary B. The permutation matrix P can always be constructed in such a way that in the final instance the blocks $A_{ii}, i = 1, \ldots, r$ are irreducible.

To test when this form is achieved we apply the theory already developed. The matrices A_{ii} are irreducible if, and only if, the digraphs of the matrices A_{ii} are strictly connected $(i = 1, \ldots, r)$.

The eigenvalues of A are of course the eigenvalues of the diagonal blocks A_{ii}. The inverse of A can likewise be calculated from the inverse of the diagonal blocks [45]. Permutation algorithms for isolating the strictly connected components of the digraph are constructed by the *achievement matrix*.

Similar considerations are applicable to block diagonalization.
$$PAP^{-1} := \begin{bmatrix} A_{11} & & 0 \\ & \ddots & \\ 0 & & A_{rr} \end{bmatrix} \tag{7.21}$$

Eigenvalues of A are again eigenvalues of the matrices A_{ii}. All A_{ii} are irreducible.

8 Eigenvalues and eigenvectors

Large eigenvalue problems arise in many scientific and technical problems. Normally such problems are sparse and the matrices encountered fall roughly into 3 classes:

- structured matrices

- modifications of structured matrices

- unstructured matrices.

Highly structured matrices arise in the methods of finite differences and finite elements [55]. These methods are appropriate for solving problems in continuum mechanics. Much work has been invested in the efficient calculation of solutions of linear sparse systems which make use of such matrices. In addition, these algorithms are used for solving eigenvalue problems.

Another class of structured sparse matrices is that of band matrices with small band width. In this context the reader is referred to [59], [60]. In many problems the matrix has no special structure; but this does not mean that the elements of the matrix are distributed by chance. Rather more the distribution of the elements reflects the underlying physical problem which is not completely regular.

Many well-known publications in the field of large eigenvalue problems are dedicated to the general eigenvalue problem $Ax = \lambda Bx$ [30]. In this case the matrices A and B are symmetric and B is generally positive definite; moreover, all eigenvalues are real. The numerical treatment of such generalized problems is rendered more difficult by the fact that there is no analogue to the well-tried method of power iteration [1] which is the basis of many algorithms for solving the conventional eigenvalue problem $Ax = \lambda x$.

Many of the well-known methods deal with eigenvalue problems involving symmetric matrices A or have not yet been translated to the nonsymmetric problem. There are several reasons for this. One of these is that the non-symmetric problem is considerably more difficult than the symmetric problem. In this connection defective matrices cause extreme trouble [7]. On the other hand, many large eigenvalue problems are symmetric by nature for which reason a good deal of work has been done in this field.

This Chapter will give no more than a brief extract from the multitude of well-tried algorithms in the form of a condensed survey. For details and comprehensive representations of the algorithms the reader is referred to the bibliography.

The general algebraic eigenvalue problem is equivalent to the determination of the roots of the *characteristic polynomial* $\varphi_A(t)$ of A:

$$\varphi_A(t) := det(A - tI) = (-1)^n[t^n + c_{n-1}t^{n-1} + \ldots + c_1 t + c_0]; \qquad c_i \in I\!K \qquad (8.1)$$

The zeros of $\varphi_A(t)$ are precisely the eigenvalues of A, and if λ is an eigenvalue of A, then

$$\varphi_A(\lambda) = 0. \qquad (8.2)$$

Generally the algorithms used for solving (8.2) are iterative.

Most of the well-tried algorithms for small and dense eigenvalue problems transform the given matrix at the first step into a condensed form. These direct reductions start with the given matrix $A_0 := A$ and produce a sequence of similarity transformations

$$A_{k+1} := P_{k+1}^{-1} A_k P_{k+1} \qquad (8.3)$$

so that for $k = n$ the matrix A_n has a special structure; for example, A_n is tridiagonal or A_n is of *Hessenberg form* [1]:

$$A_n := \begin{bmatrix} & \diagdown & 0 \\ & \diagdown & \\ 0 & & \diagdown \end{bmatrix} \quad \text{or} \quad A_n := \begin{bmatrix} & \diagdown & \\ & \diagdown & \\ 0 & & \diagdown \end{bmatrix}.$$

If a large sparse matrix can be represented in the fast storage of a computer, the given matrix A may be reduced by (8.3) to a condensed form; for example, a symmetric matrix $A = A^T$ can be transformed into a tridiagonal matrix by using *Householder transformations*, i.e. transformations of the form

$$P := I - 2ww^T, \quad \text{with } P \in I\!R^{n \times n}, \ w \in I\!R^n, \ \text{and } w^T w = 1. \qquad (8.4)$$

In order to transform a real $(n \times n)$ matrix into Hessenberg form, $n-2$ Householder transformations P_i are determined so that $Q := P_1 \ldots P_{n-2}$ is orthogonal and $Q^T A Q$ is of Hessenberg form.

Suppose that

$$A_k := P_k A_{k-1} P_k, \quad 1 \le k \le n-1, \quad A_0 = A. \tag{8.5}$$

The matrices P_k are determined in such a way that the transformation with P_k cancels the elements of the kth column below the subdiagonal; in this way the zeros created in the previous steps remain unaltered. Before executing the kth step the matrix A has been reduced to the following form:

$$A_{k-1} := \left[\begin{array}{c|cc} H_{k-1} & C_{k-1} \\ \hline 0 & b_{k-1} & B_{k-1} \end{array} \right] \begin{array}{l} \}k \\ \}n-k. \end{array}$$

Here H_{k-1} is a square Hessenberg matrix of order k, B_{k-1} is a square matrix of order $n-k$, and b_{k-1} is a vector having $n-k$ components. Then for A_k we have [1]:

$$A_k := P_k A_{k-1} P_k = \left[\begin{array}{c|cc} H_{k-1} & C_{k-1} Q_k \\ \hline 0 & \tilde{b}_{k-1} & Q_k B_{k-1} Q_k \end{array} \right],$$

where

$$\tilde{b}_{k-1} := Q_k b_{k-1}$$

and

$$P_k := \left[\begin{array}{c|c} I & 0 \\ \hline 0 & Q_k \end{array} \right] \begin{array}{l} \}k \\ \}n-k. \end{array}$$

This transformation does not change the matrix H_{k-1}.

The main enemy of such transformations is the fill-in, which is relatively larger for eigenvalue problems than for linear systems. This is due to two essential reasons. Similarity transformations of the form (8.3) are more complicated than the one-sided transformations for solving linear systems. On the other hand, orthogonal transformations are used for symmetric matrices to preserve symmetry; these transformations tend towards a larger fill-in than the non-orthogonal transformations. A skilful choice of pivot during the reduction may simplify the fill-in problem [57].

When the matrix has been transformed into condensed form and the eigenvalues are to be calculated, a suitable algorithm for small dense matrices is the QR-algorithm:

Sketch of the QR-algorithm

Let $A_1 \in \mathbb{C}^{n\times n}$ be a square matrix. The QR-algorithm produces a sequence $\{A_i\}, i \in \mathbb{N}$, of square matrices as follows:

First, for $i \in \mathbb{N}$, the matrix A_i is decomposed into a unitary matrix Q_i, i.e. $Q_i^H Q_i = I$, and a right-hand triangular matrix R_i:

$$A_i = Q_i R_i ; \tag{8.6}$$

then A_{i+1} is computed from

$$A_{i+1} = R_i Q_i. \tag{8.7}$$

A_{i+1} is unitarily similar to A_i, for

$$A_{i+1} = R_i Q_i = Q_i^H A_i Q_i \tag{8.8}$$

holds, where $R_i = Q_i^H A_i$.

This sequence of similar matrices is the basis of one of the most efficient algorithms for determining all eigenvalues of a matrix. Subject to the fulfillment of certain preconditions the sequence converges to a right-hand triangular matrix or, in general, to a quasi right-hand triangular matrix whose diagonal elements permit the calculations of the eigenvalues of the matrix [54]. Here different cases have to be distinguished.

For large sparse matrices the QR-algorithm causes some trouble. In general the transformations of the QR-algorithm destroy the sparseness of the matrix. Hence the use of the QR-algorithm is not a good choice for a sparse Hessenberg matrix. It is well suited for determining all eigenvalues of a symmetric tridiagonal matrix, but not for the calculation of only some eigenvalues. Finally, if eigenvectors are to be determined either the original matrix or the QR-transformations have to be preserved.

In connection with the general eigenvalue problem

$$Ax = \lambda Bx, \tag{8.9}$$

where A is symmetric and B is positive definite, many algorithms are based on the fact that *Rayleigh's quotient*

$$R[x] := \frac{x^T A x}{x^T B x}, \quad x \neq 0, \tag{8.10}$$

is minimal for the smallest eigenvalue λ_1 of the problem. This minimum is attained if x is an eigenvector of the eigenvalue λ_1.

The idea is to generate a sequence $\{x^{(k)}\}_{k\in N}$ of vectors, where

$$x^{(k+1)} := x^{(k)} + \alpha_k q_k ; \tag{8.11}$$

α_k is a scalar and q_k is a correction vector. The elements α_k and q_k are chosen in such a way that the sequence $\{R[x^{(k)}]\}$ decreases and so converges towards a limit point.

In the special case $B = I$ these algorithms reduce to algorithms for the solution of the ordinary symmetric eigenvalue problem.

The simplest form of this method is the *relaxation method*. Here q_k is chosen as one of the coordinate vectors e_i such that only the ith component of $x^{(k)}$ is changed. α_k is determined such that $x^{(k+1)}$ satisfies the following equation:

$$(a_i^T - R[x^{(k)}]b_i^T)x^{(k+1)} = 0. \tag{8.12}$$

a_i^T is the ith row of A; b_i^T is the ith row of B. This is a step of the ordinary relaxation method for solving the homogeneous systems

$$(A - R[x^{(k)}]B)x = 0. \tag{8.13}$$

The different versions of this algorithm used in computers choose cyclically the direction vectors among the set of coordinate vectors e_k, $k = 1, \ldots, n$. The relaxation method permits an overrelaxation such that α_k is chosen larger than necessary to solve (8.12). This guarantees in some cases an acceleration of convergence.

The method of *coordinate overrelaxation* is also intended to find the smallest eigenvalue of the general eigenvalue problem

$$(A - \lambda B)x = 0; \tag{8.14}$$

A and B are symmetric and sparse and B is positive definite. In this connection the smallest eigenvalue is obtained by the smallest stationary values of Rayleigh's quotient

$$R[x] = \frac{(x, Ax)}{(x, Bx)}, \quad x \neq 0. \tag{8.15}$$

Let λ_i denote the different eigenvalues of (8.14) with

$$\lambda_1 < \lambda_2 < \lambda_3 < \ldots . \tag{8.16}$$

The multiplicity of an eigenvalue can be larger than one. In the following only the problem of finding the smallest eigenvalue λ_1 is dealt with, i.e. the minimum of (8.15) must be calculated.

First the algorithm of *coordinate relaxation* is sketched.

To find the minimum of (8.15) one starts with an arbitrary vector $x \neq 0$, satisfying the inequality

$$R[x] = r \geq \lambda_1. \tag{8.17}$$

In order to reduce the actual value r the vector x is changed only in its kth component; i.e. the minimum $R[x]$ in the set of vectors $\{\bar{x} \mid \bar{x} = ax + be_k\}$ is to be determined. a and b are scalars which do not vanish simultaneously. So the minimum of

$$R[\bar{x}] = \frac{(ax + be_k, A(ax + be_k))}{(ax + be_k, B(ax + be_k))} \tag{8.18}$$

$$= \frac{\alpha a^2 + 2fab + pb^2}{\beta a^2 + 2gab + qb^2} \tag{8.19}$$

must be found, where:

$$\begin{aligned} \alpha &:= (x, Ax) \\ \beta &:= (x, Bx) \end{aligned} \tag{8.20}$$

$$\begin{aligned} f &:= (e_k, Ax) = u_k \\ g &:= (e_k, Bx) = v_k \end{aligned} \tag{8.21}$$

$$\begin{aligned} u &:= Ax \\ v &:= Bx \end{aligned} \tag{8.22}$$

$$\begin{aligned} p &:= (e_k, Ae_k) = a_{kk} \\ q &:= (e_k, Be_k) = b_{kk}. \end{aligned} \tag{8.23}$$

u_k and v_k denote the kth component of the vectors u and v respectively, defined in (8.22).

The problem of minimizing (8.18) leads to the small-scale general eigenvalue problem

$$\begin{aligned} (\alpha - \acute{r}\beta)a + (u_k - \acute{r}v_k)b &= 0 \\ (u_k - \acute{r}v_k)a + (a_{kk} - \acute{r}b_{kk})b &= 0. \end{aligned} \tag{8.24}$$

The minimum of $R[\bar{x}]$ is equal to the smaller of the two eigenvalues $\acute{r}$ and the scalars a and b are the components of the corresponding eigenvector.

Since an eigenvector is determined only up to a multiplicative constant, one of its components can be set equal to unity. Normally we set $a = 1$. Further cases are distinguished in [51].

Now a single step of coordinate relaxation can be sketched.

For a systematic decrease of the Rayleigh quotients the unit vectors are chosen cyclically; the sequence of n single steps is called a *cycle*. Note that α and β can be computed recursively. Finally the value of b, which is normally obtained from (8.25), is multiplied by a constant relaxation factor ω and the following *algorithm of coordinate overrelaxation* is obtained:

$$\begin{aligned} start: \quad & \text{choose } x \neq 0 \\ & \alpha := (x, Ax), \\ & \beta := (x, Bx), \\ & r := \tfrac{\alpha}{\beta} \end{aligned}$$

$$\begin{aligned} cycle: \quad & \text{for } i = 1, \ldots, n \text{ calculate:} \\ & u_i := \sum_{k=1}^{n} a_{ik} x_k, \\ & v_i := \sum_{k=1}^{n} b_{ik} x_k. \end{aligned}$$

From (8.24) and (8.25) calculate $\acute{r} \leq r$.

Different cases:

(i) If $|a_{ii} - \acute{r}b_{ii}| > \epsilon$, then

$$\begin{aligned} b &:= -\omega(u_i - \acute{r}v_i)/(a_{ii} - \acute{r}b_{ii}) \\ x_i &:= x_i + b \\ \alpha &:= \alpha + 2u_i b + a_{ii}b^2 \\ \beta &:= \beta + 2v_i b + b_{ii}b^2 \\ r &:= \tfrac{\alpha}{\beta}. \end{aligned}$$

(ii) If $|a_{ii} - \acute{r}b_{ii}| \leq \epsilon$ and $|\alpha - \acute{r}\beta| > \epsilon$, then

$$\begin{aligned} x &:= e_i \\ \alpha &:= a_{ii} \\ \beta &:= b_{ii} \\ r &:= \tfrac{\alpha}{\beta}. \end{aligned}$$

(iii) If $|a_{ii} - \acute{r}b_{ii}| \leq \epsilon$ and $|\alpha - \acute{r}\beta| \leq \epsilon$, then

$$x, \alpha, \beta, r \text{ remain unchanged.}$$

The sparseness of the matrices A and B enters the calculation of the values v_i and u_i. ϵ is a suitably chosen tolerance.

The cycles of the algorithms are repeated until the value of the Rayleigh quotient becomes stationary over a complete cycle and the maximum absolute value of the b-values is smaller than a given bound. In case of normed vectors x this tolerance is, in a certain sense, a measure of the exactness of the approximation of the eigenvector.

The convergence of the coordinate relaxation ($\omega = 1$) is embodied in the following Theorem [51]:

Theorem 8.26
The sequence of the vectors $x^{(k)}$, determined by the algorithm of coordinate relaxation, converges towards the direction of the eigenvector x_1 corresponding to the smallest eigenvalue λ_1, if

- λ_1 is simple;
- $R[x^{(0)}] < \lambda_2$, $x^{(0)}$ denoting the commencing vector;
- $R[x^{(0)}] < \min_i(\tfrac{a_{ii}}{b_{ii}})$.

Further details concerning coordinate relaxation can be found in [51].

Various algorithms for the calculation of an approximate eigenvector of a matrix A are obained by the so-called *Krylow-sequences* $x, Ax, A^2x, \ldots$. The vectors

$$x^{(i+1)} := Ax^{(i)}, \quad i = 0, 1, 2, \ldots, \quad x^{(0)} \in \mathbb{R}^n \qquad (8.27)$$

are called also *iterated vectors*. Since the iterated vectors $x^{(i)}$ are generated by matrix vector multiplications they are very appropriate for large sparse problems. A great many of these algorithms treat only the case of symmetric matrices.

In this context one of the oldest methods for calculating an eigenvector is the *power iteration* (method of v. Mises). It is based on the fact that the iterated vectors (8.27) converge to an eigenvector of A.

Let $\lambda_1, \lambda_2, \ldots, \lambda_n$, be the eigenvalues of A, where

$$|\lambda_1| > |\lambda_2| \geq |\lambda_3| \geq \ldots \geq |\lambda_n|. \tag{8.28}$$

Under weak conditions imposed on $x^{(0)} \in I\!R^n$ the sequence $A^i x^{(0)}$ converges to the eigenvector z_1 associated with λ_1. The convergence is essentially linear with λ_2/λ_1. To avoid many matrix multiplications, $A^i x^{(0)}$ is calculated iteratively by

$$x^{(k+1)} := A x^{(k)}.$$

Power iteration converges badly if no strictly dominant eigenvalue exists. A remedial measure is the transformation of A so that the transformed matrix has a strictly dominant eigenvalue. This is the *shifted power iteration*. By shifting the spectrum, a matrix $B = A - \alpha I$, $\alpha \in I\!R$, is obtained having the same eigenvectors as A and the eigenvalues $\mu_k = \lambda_k - \alpha$. If A is symmetric with eigenvalues $\lambda_1 > \lambda_2 > \ldots > \lambda_n$, the optimal value of α is

$$\alpha = \frac{1}{2}(\lambda_2 + \lambda_n).$$

Clearly this value of α can be calculated only if estimates for λ_2 and λ_n exist. In the case of a positive definite matrix A the eigenvalue λ_n may be estimated by zero. From the eigenvectors the corresponding eigenvalues can be calculated. This method has been extended to the *simultaneous iteration* of several eigenvectors and their eigenvalues [43].

The reader is referred to the literature cited for additional algorithms.

9 Parallel numerical algorithms

The effective treatment of large sparse systems makes it desirable to develop new, fast and efficient methods. In this context parallel algorithms are of increasing importance.

In general, numerical algorithms can be classified in different ways, for example as algebraic or analytic algorithms, as finite or infinite algorithms, or as direct or iterative algorithms. In the last few years a new classification has become more important: that is, whether an algorithm is serial or parallel. The difference between such algorithms has become highly significant due to the development of parallel and pipeline computers. These machines permit the parallel execution of arithmetical operations. They are able to handle a great deal of information and there are often hardware possibilities on different levels.

The basic idea of such a parallel computer is that programs using n processors should be n times faster than programs using only one processor. But experience and theory show that the real speed-up is less.

There is need for a *methodology of parallel algorithms* aimed at the optimal use of parallel computers. Another more theoretical question is how to solve problems using maximal parallelism.

The development of high-speed computers makes it necessary to reconfigure well-known methods for solving large and complex systems and to develop new efficient algorithms.

The structure of these algorithms and their software are inherently dependent on the architecture of the computer system used, and vice versa.

The following Figure 9.1 gives an idea of the importance of adapting the methods and the architecture of the computer system to the problems under consideration.

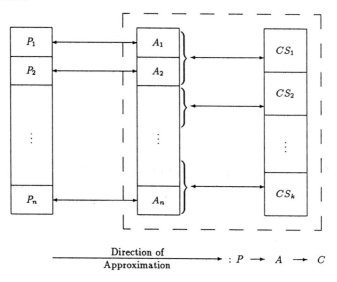

$$: P \rightarrow A \rightarrow C$$

P : problem
P_i : subproblems
A_i : subalgorithms
CS_k : computer systems

$A := [A_1, A_2, \ldots, A_n]$

$C := [CS_1, \ldots, CS_k]$

Fig. 9.1 – Adaptive methods

Among currently significant applications are: curve fitting, weather forecasting, spin model, simplex optimization, physical field evaluation, evaluation of P/N transitions, solution of systems of linear equations, structure analysis, image processing and others.

Characteristics of parallel computers

A well-known classification of computer systems is that of Flynn [16]:

- *SISD* machine : *S*ingle-*I*nstruction-*S*ingle-*D*ata-stream machine
  (this is the conventional von Neumann machine)
- *SIMD* machine : *S*ingle-*I*nstruction-*M*ultiple-*D*ata-stream machine
- *MISD* machine : *M*ultiple-*I*nstruction-*S*ingle-*D*ata-stream machine
- *MIMD* machine: *M*ultiple-*I*nstruction-*M*ultiple-*D*ata-stream machine

The development of new computer architectures with different levels of parallelism requires a detailed classification essential for the comparison of computers [29]. Figure 9.2 shows a general model of a parallel computer:

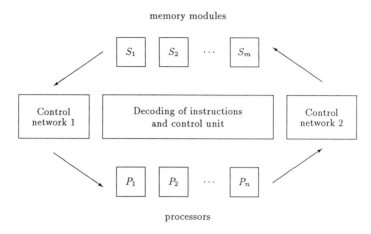

Fig. 9.2 – General configuration with different levels of parallelism

Parallelism is possible

- within the control unit
- among the processors
- among the stores
- in the data central network.

Modern computers include:

- CYBER 203/205 ($SIMD$ machine)
- CRAY 1s, CRAY X-MP, CRAY 2
- HEP-Denelcor ($MIMD$ machine)
- Hitachi S9/IAP (integrated array processor) ($MIMD$ machine)
- Alliant
- FPS (Floating Point System).

For some further details see [49].

9.1 Basic concepts for the development of parallel algorithms

In a parallel algorithm, because more than one task module can be executed at a time, concurrency control is needed to ensure the correctness of the concurrent execution. The concurrency control enforces desired interactions among task modules so that the overall execution of the parallel algorithm will be correct. Figure 9.3 represents the space of concurrency controls that can be used in parallel algorithms.

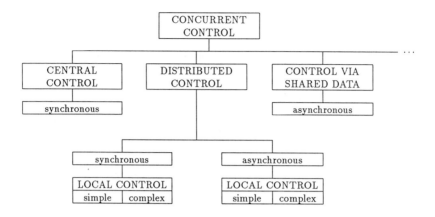

Fig. 9.3 – Concurrent controls

The leaves of the tree represent various types of concurrency controls. Concurrent control has a great influence on the structure of an algorithm.

Table 9.1 Characterization of algorithms

Type	Concurrent control	Remarks
$SIMD$ algorithm	central control unit - $SIMD$ -	$SIMD$ machines correspond to synchronous algorithms that require central control units
$MIMD$ algorithm	asynchronous, shared memory - $MIMD$ -	$MIMD$ machines correspond to asynchronous algorithms with relatively large granularities
systolic algorithm	distributed control achieved by simple local control	LSI and VLSI machines [3] for special algorithms

---

[3] LSI: Large Scale Integrated; VLSI: Very Large Scale Integrated

Note 9.1

- Systolic algorithms are designed for direct hardware implementation [32].
- $MIMD$ algorithms are designed to be executed on general purpose multiprocessors.
- $SIMD$ algorithms lie between the two other types.

Table 9.2 Examples of applications in parallel algorithm space

Application	Examples
$SIMD$ algorithm	numerical relaxation for image processing partial differential equations, Gaussian elimination with pivoting
$MIMD$ algorithm	concurrent data base algorithms (concurrent accesses to binary search trees), chaotic relaxation dynamic scheduling algorithms, algorithms with large module granularities
systolic algorithm using	
- one-dimensional linear arrays	discrete Fourier transform (DFT), solution of triangular linear systems, recurrence evaluation
- two-dimensional square arrays	dynamic programming, image processing numerical relaxation, graph algorithms

Note 9.2
In a *one-dimensional linear array machine* data can flow simultaneously in both directions depending on the algorithm.

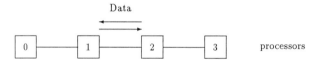

Fig. 9.4 – One-dimensional array

In a *two-dimensional array machine* the processors are distributed in a two-dimensional array so that the connections are symmetrical and all of the same length. They have to cover the whole array. There are three figures having such properties:

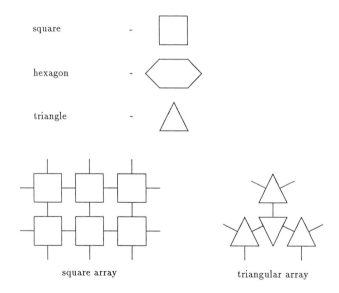

Fig. 9.5 – Two-dimensional arrays

9.2 Arithmetical expressions

An algorithm **A** is represented as a composition of arithmetical expressions E_k

$$\mathbf{A} = E_1 \circ E_2 \circ \ldots \circ E_n. \qquad (9.3)$$

The problem is to find the optimal code for the arithmetical expression E_k in a serial or parallel computer. The generation of a minimal code for such problem is called an *np-complete problem* or *node-parse-complete problem*. It is possible to find a minimal evaluation of arithmetical expressions with common subexpressions without using associativity and commutativity.

9.2 Arithmetical expressions

Example 9.4
Consider the given expressions

$$E_1 := a_3 \cdot (a_4 + a_5),$$
$$E_2 := a_1 \cdot a_2 + ((a_3 \cdot (a_4 + a_5) + a_3 \cdot (a_4 + a_5))$$
$$- (a_3 \cdot (a_4 + a_5) + (a_4 + a_5))),$$
$$E_3 := (a_3 \cdot (a_4 + a_5) + a_3 \cdot (a_4 + a_5)) - (a_3 \cdot (a_4 + a_5)$$
$$+ (a_4 + a_5))/(a_1 \cdot a_2).$$

Then there exist the following subexpressions:

$$T_1 := (+, a_4, a_5) \quad ; \quad E_1 := (\cdot, a_3, T_1)$$
$$T_2 := (-, T_4, T_5) \quad ; \quad T_3 := (\cdot, a_1, a_2).$$

The additional subexpressions T_4 and T_5 are illustrated by Figure 9.6.

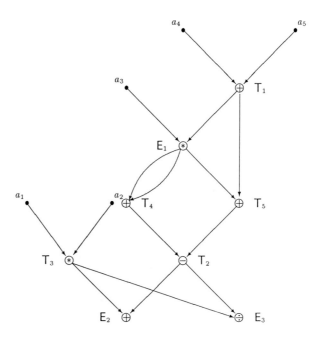

Fig. 9.6 – Example of minimal evaluation

Generally the cost of solving np-complete problems is exponential. On the other hand it can be shown that the so-called *feed-back-node problems* of the np-class can be solved at polynomial cost when treated as an enumeration problem [2]. Other authors try to solve this problem of optimal node generation by using directed, acyclic graphs - so-called *dags*.

Algorithms for the parallel evaluation of arithmetical expressions are based on the construction of equivalent arithmetical expressions.

Definition 9.5
Two arithmetical expressions E and Ẽ are said to be *equivalent* if it is possible to pass from E to Ẽ by a finite number of applications of commutative, distributive or associative laws.

Example 9.6
The following two arithmetical expressions E_1 and $\tilde{E}_1$ are equivalent ($E_1 \sim \tilde{E}_1$):

$$E_1 := ((a_1 \cdot a_2 + a_3) \cdot a_4 + a_5) \cdot a_6 + a_7$$
$$\tilde{E}_1 := (a_1 \cdot a_2 \cdot a_4 \cdot a_6 + a_3 \cdot a_4 \cdot a_6) + (a_5 \cdot a_6 + a_7).$$

E_1 is called a *generalized Horner algorithm*.

In the following Figure 9.7 the dags $\mathcal{G}_1$ and $\tilde{\mathcal{G}}_1$ describe the serial and parallel evaluation of E_1 and $\tilde{E}_1$:

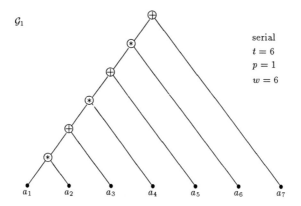

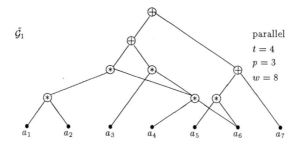

$t :=$ number of serial or parallel time steps
$p :=$ number of processors
$w :=$ number of operations performed by the algorithm

Fig. 9.7 – Generalized Horner algorithm

The dag $\mathcal{G}_1$ of E_1 is a binary tree. Under the applications of associative, commutative and distributive laws $\mathcal{G}_1$ is transformed into $\tilde{\mathcal{G}}_1$ of $\tilde{\mathsf{E}}_1$ ($\mathsf{E}_1 \sim \tilde{\mathsf{E}}_1$). The algorithm for $\mathcal{G}_1$ should minimize $\tilde{t} = \tilde{t}(p)$.

Lemma 9.7
If a computation can be performed in time t with w operations and a sufficiently large number of processors $\tilde{p}$, then the maximal computation time, according to [3], is
$$\tilde{t} := \frac{t + (w - t)}{\tilde{p}}.$$

Theorem 9.8
Let **A** be any algorithm for evaluation of the generalized Horner algorithm
$$\mathsf{H} := (\ldots (a_1 \cdot a_2 + a_3) \cdot a_4 + a_5) \cdot a_6 + \ldots) \cdot a_{2n} + a_{2n+1}; \qquad (9.9)$$
then
$$w \geq 3n - t/2.$$
This is equivalent to $t < 2n \Rightarrow w > 2n$, [24].

Note 9.10
For a hypothetical $\dot{MIMD}$ machine it is assumed that

- $op \in \mathcal{M} := \{+, -, *, /\}$ can be performed by each processor
- different processors perform different operations $op \in \mathcal{M}$
- $op \notin \mathcal{M}$ requires no time
- $op \in \mathcal{M}$ requires one time unit.

9.3 Some remarks on the development of parallel algorithms

The previous considerations show that the kind of parallelism of the computer, namely whether it is an $SIMD$ or $MIMD$ machine, essentially influences the nature of the algorithm. Moreover, the arrangement and dynamic sorting of data in memory, to which the algorithm needs to have parallel access, is of importance.

The basis of numerical methods is the evaluation of arithmetical expressions. The path following by the evaluation can be represented by graphs (trees). Application of the laws of real numbers often leads to *tree height reduction* whose possible utilization is dependent on the type of computer considered.

One way in which arithmetical expressions can be parallelized is illustrated by the simple expression
$$\mathsf{E} := a_4 + (a_3 + (a_2 + a_1)), \quad a_i \in \mathbb{R}, \; i = 1, \ldots, 4, \qquad (9.11)$$
which, by utilization of the associativity of addition, can be transformed into
$$\tilde{\mathsf{E}} := (a_4 + a_3) + (a_2 + a_1), \quad a_i \in \mathbb{R}, \; i = 1, \ldots, 4; \qquad (9.12)$$
this means that two additions can be made in parallel.

Given an arbitrary expression E_0 one tries to split this into two smaller expressions E_1 and E_2, which can be calculated simultaneously, each by one processor. For the execution on a $SIMD$ machine the following conditions must be valid:

- A function f exists with $E_0 = f(E_1, E_2)$.
- E_1 and E_2 are computed independently of each other and are of the same complexity.
- E_1 and E_2 require the same series of computation.

Further splitting of E_1 and E_2 according to these conditions leads to E_0 by *recursive doubling* [27]. To describe this method consider a set

$$S = \{a_1, a_2, \ldots, a_N | N = 2^k,\ k \in \mathbb{N}\} \subset \mathbb{R}$$

and an associative operation $op \in \mathcal{M} := \{+, *, \max, \ldots\}$ in S. Now the expression

$$E = a_1\ op\ a_2\ op\ \ldots\ op\ a_N$$

can be calculated.

Example 9.13
Consider the expression

$$E = a_1 + a_2 + a_3 + a_4.$$

By using the associative property of addition we get

$$\tilde{E} = (a_1 + a_2) + (a_3 + a_4).$$

The following Figure 9.8 shows the comparison of these both expressions:

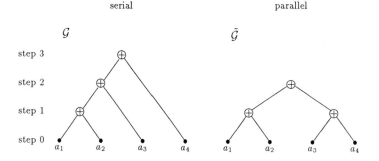

Fig. 9.8 – $SIMD$ structure

Generally, recursive doubling with $N = 2^n$ elements requires $\log_2 N$ parallel steps. Serial implementation requires $N - 1$ steps.

Example 9.14

Suppose that it is required to compute

$$E = a_1 + a_2 * a_3 + a_4.$$

The parsetree $\mathcal{G}$ of E (Figure 9.9) is not a unique tree and no tree height reduction can be obtained by applying the associative law. By using the commutative property of addition, E is transformed into

$$\tilde{E} = (a_1 + a_4) + a_2 * a_3$$

and the following parsetrees are obtained:

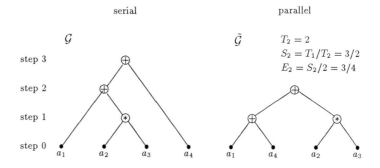

S: speed-up; T: number of time units; E: efficiency

Fig. 9.9 - $MIMD$ structure

The following questions arise among others :

- How many tree height reductions can be achieved for a given arithmetical expression?

- Can general algorithms be given for tree height reduction?

- How many processors are needed for optimality, in some sense?

For more details see [49].

9.4 Recurrence relations

Recurrence relations are appropriate to problems whose solutions are expressed in the form of a sequence $x_1, x_2, \ldots, x_n$, $x_i \in \mathbb{R}$, where each x_i, $i = 1, 2, \ldots, n$ may depend on x_j, $j < i$.

A general example of recurrence relations is a time-dependent linear system in which the state x_t of the system at time t is a linear function of the state at time $t-1$:

$$
\begin{aligned}
x_1 &:= c_1 & & \text{initial condition} \\
x_2 &:= a_2 x_1 + c_2 \\
&\vdots \\
x_t &:= a_t x_{t-1} + c_t & & \text{recurrence equation} \quad (9.15) \\
&\vdots \\
x_n &:= a_n x_{n-1} + c_n.
\end{aligned}
$$

At first sight serial computation seems to be possible.

Definition 9.16
An mth order *linear recurrence system* $R<n,m>$ for n equations is defined for $m \leq n-1$ by

$$R<n,m>: \quad x_k := \begin{cases} 0, & \text{for } k \leq 0 \\ c_k + \sum_{j=k-m}^{k-1} a_{kj} x_j, & \text{for } 1 \leq k \leq n. \end{cases} \quad (9.17)$$

If $m = n-1$ this system is called an *ordinary linear system of recurrence equations* and denoted by $R<n>$.

Example 9.18
The inner product of two vectors

$$x = (x_1, x_2, \ldots, x_n) \text{ and } y = (y_1, y_2, \ldots, y_n)$$

can be written as a linear recurrence equation of first order

$$z := z + x_k y_k, \quad 1 \leq k \leq n$$

with the initial condition $z := 0$.

Example 9.19
The Fibonacci sequence

$$
\begin{aligned}
f_k &:= f_{k-1} + f_{k-2}, \quad 3 \leq k \leq n, \\
f_1 &:= f_2 := 1
\end{aligned}
$$

may be considered as a second order recurrence equation.

Using matrix-vector notation (9.17) can be written as

$$x = c + Ax,$$

where

$$x = (x_1, \ldots, x_n) \text{ and } c = (c_1, \ldots, c_n)$$

and

$$A = [a_{ik}], \quad i, k = 1, 2, \ldots, n$$

is a strictly lower triangular matrix with $a_{ik} = 0$ for $i \leq k$ or $i - k \geq m$. A is a band matrix for $m < n-1$.

9.4 Recurrence relations

Example 9.20
For $n = 4$ we have the following ordinary recurrence system $R<4>$:

$$\begin{aligned} x_1 &= c_1 \\ x_2 &= c_2 + a_{21}x_1 \\ x_3 &= c_3 + a_{31}x_1 + a_{32}x_2 \\ x_4 &= c_4 + a_{41}x_1 + a_{42}x_2 + a_{43}x_3. \end{aligned}$$

Definition 9.21
A general mth order *recurrence system* $R<n,m>$ is defined by

$$R<n,m>:\ x_k := H[a_k; x_{k-1}, x_{k-2}, \ldots, x_{k-m}],\quad 1 \leq k \leq n$$

with m initial conditions $x_{-m+1}, x_{-m+2}, \ldots, x_0$. H is called the *recursion function* and a_k is a vector of parameters which are independent of the x_i.

This definition is suited to $SIMD$ machines.

Example 9.22
A simple example of a first order recurrence system is given by (9.15):

$$\begin{aligned} x_1 &:= c_1 \qquad\qquad \text{initial condition} \\ x_k &:= a_k x_{k-1} + c_k, \quad \text{for } 2 \leq k \leq n \\ &= H[a_k; x_{k-1}] \end{aligned}$$

where the parameter vector a_k is of length 2 and of the form $a_k = (a_k, c_k)$.

For a parallel solution of these systems the associativity of the recursion function H is required. However, there often exist so-called *companion functions* G associated with H, which possess associative properties.

Definition 9.23
A function G is said to be a *companion function* to the recursion function H if, for all $x \in \mathbb{R}$ and for all parameter vectors $a, b \in \mathbb{R}^p$

$$H[a; H[b; x]] = H[G(a,b); x] \qquad (9.24)$$

holds, where $G: \mathbb{R}^p \times \mathbb{R}^p \to \mathbb{R}^p$.

All companion functions have the following property:

Theorem 9.25
Every companion function G is associative with respect to its recursion function H; i.e. for all $x \in \mathbb{R}$ and $a, b, c \in \mathbb{R}^p$ we have

$$H[G(a, G(b,c)); x] = H[G(G(a,b), c); x].$$

Proof:
The proof is given in [49]. □

If such functions G can be found, the parallelization of the recurrence can be reduced to the construction of a companion function.

Example 9.26
Consider the first order recurrence
$$x_k := H[a_k; x_{k-1}]$$
where $a_k \in \mathbb{R}^p$ and x_0 is the initial value.

$$
\begin{aligned}
x_2 &= H[a_2; x_1] \\
&= H[a_2; H[a_1; x_0]] \\
&= H[G(a_2, a_1); x_0], \quad \text{where } G \text{ is the companion function}
\end{aligned}
$$

$$
\begin{aligned}
x_4 &= H[a_4; x_3] \\
&= H[a_4; H[a_3; x_2]] \\
&= H[G(a_4, a_3); x_2] \\
&= H[G(a_4, a_3); H[G(a_2, a_1); x_0]] \\
&= H[G(G(a_4, a_3), G(a_2, a_1)); x_0]
\end{aligned}
$$

$$
\begin{aligned}
x_8 &= H[a_8 : x_7] \\
&= H[a_8; H[a_7; H[a_6; H[a_5; x_4]]]] \\
&= H[G(a_8, a_7); H[G(a_6, a_5); x_4]] \\
&= H[G(a_8, a_7); H[G(a_6, a_5); H[G(G(a_4, a_3), G(a_2, a_1); x_0)]]] \\
&= H[G(G(a_8, a_7), G(a_6, a_5)); H[G(G(a_4, a_3), G(a_2, a_1); x_0)]] \\
&= H[G(G(\underbrace{G(a_8, a_7), G(a_6, a_5)}), G(\underbrace{G(a_4, a_3), G(a_2, a_1)})); x_0]
\end{aligned}
$$

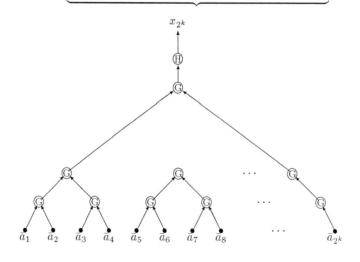

Fig. 9.10 – Parsetree for first order recurrence

9.5 Parallel solution of linear systems on SIMD machines

The *column-sweep algorithm* serves for solving an $R<n>$ system of recurrence relations by a fast and efficient method such that with $N = O(n)$ processors the system can be computed in $O(n)$ time unit steps. Referring back to the set of equations of Example 9.20 we get the following algorithm:

Step 1 :
x_1 is known; then the expressions

$$c_i^{(1)} := a_{i1}x_1 + c_i, \quad i = 1, \ldots, n$$

can be calculated in parallel. Now x_2 is known and there remains an $R<n-1>$ system.
Step 2 :
x_2 is known; the expressions

$$c_i^{(2)} := a_{i2}x_2 + c_i^{(1)} \text{ with } c_i^{(1)} := a_{i1}x_1 + c_i$$

are calculated in parallel, and then x_3 is known. This procedure continues and in general we have:
Step k :
x_k is known and the expressions

$$c_i^{(k)} := a_{ik}x_k + c_i^{(k-1)} \text{ with } c_i^{(k-1)} := a_{i,k-1}x_{k-1} + c_i^{(k-2)}$$

are calculated, giving x_{k+1}. Continuing thus we arrive at ...
Step $n-1$:
The calculation of x_n completes the calculation of the vector x.

This algorithm requires $n-1$ processors at step 1 and fewer thereafter.

In applications, $R<n,m>$ systems with $m \ll n$ are of interest. The solution of such system by the column-sweep algorithm requires at most m processors. So, if a large number of processors is available the procedure is disadvantageous[49].

One of the fastest procedures for the solution of an $R<n,m>$ system having small m is the so-called *recurrent product form algorithm*. The algorithm is developed from the product form representation of the solution of the $R<n>$ system

$$x = Ax + c, \tag{9.27}$$

where A is a strictly lower triangular matrix. The solution of (9.27) is

$$x = (I-A)^{-1}c = L^{-1}c. \tag{9.28}$$

As shown in [27] L^{-1} can be expressed in the product form

$$L^{-1} = M_n \cdot M_{n-1} \cdot \ldots \cdot M_1 \tag{9.29}$$

with

$$M_i = \begin{bmatrix} 1 & & & & & & & \\ & \ddots & & & & & & 0 \\ & & 1 & & & & & \\ & & & \frac{1}{a_{ii}} & & & & \\ & & & \frac{-a_{i+1,i}}{a_{ii}} & 1 & & & \\ & 0 & & \vdots & & \ddots & \\ & & & \frac{-a_{ni}}{a_{ii}} & & & 1 \end{bmatrix}$$

and in this case $a_{ii} = 1$ throughout. The solution vector

$$x = M_{n-1} \cdot M_{n-2} \cdot \ldots \cdot M_1 \cdot c$$

can be calculated in parallel mode by use of the recursive doubling procedure in $O(\log_2 n)$ time steps[29].

Example 9.30
Consider the $R < 4, 2 >$ system with the recurrence relations

$$\begin{array}{rcl} x_1 & = & c_1 \\ x_2 & = & c_2 + a_{21}x_1 \\ x_3 & = & c_3 + a_{31}x_1 + a_{32}x_2 \\ x_4 & = & c_4 + a_{42}x_2 + a_{43}x_3 \end{array}$$

or

$$x = c + Ax,$$

where

$$A := \begin{bmatrix} 0 & 0 & 0 & 0 \\ a_{21} & 0 & 0 & 0 \\ a_{31} & a_{32} & 0 & 0 \\ 0 & a_{42} & a_{43} & 0 \end{bmatrix}.$$

Thus

$$L := \begin{bmatrix} 1 & 0 & 0 & 0 \\ -a_{21} & 1 & 0 & 0 \\ -a_{31} & -a_{32} & 1 & 0 \\ 0 & -a_{42} & -a_{43} & 1 \end{bmatrix}.$$

Hence

$$x = L^{-1}c = M_3 M_2 M_1 c$$

$$= \begin{bmatrix} 1 & 0 & 0 & 0 \\ 0 & 1 & 0 & 0 \\ 0 & 0 & 1 & 0 \\ 0 & 0 & a_{43} & 1 \end{bmatrix} \begin{bmatrix} 1 & 0 & 0 & 0 \\ 0 & 1 & 0 & 0 \\ 0 & 0 & a_{32} & 0 \\ 0 & 0 & a_{42} & 1 \end{bmatrix} \begin{bmatrix} 1 & 0 & 0 & 0 \\ a_{21} & 1 & 0 & 0 \\ a_{31} & 0 & 1 & 0 \\ 0 & 0 & 0 & 1 \end{bmatrix} \begin{bmatrix} c_1 \\ c_2 \\ c_3 \\ c_4 \end{bmatrix}$$

$$= \begin{bmatrix} 1 & 0 & 0 & 0 \\ 0 & 1 & 0 & 0 \\ 0 & a_{32} & 1 & 0 \\ 0 & (a_{43}a_{32} + a_{42}) & a_{43} & 1 \end{bmatrix} \begin{bmatrix} c_1 \\ a_{21}c_1 + c_2 \\ a_{31}c_1 + c_3 \\ c_4 \end{bmatrix}$$

giving

$$x = \begin{bmatrix} c_1 \\ a_{21}c_1 + c_2 \\ a_{32}(a_{21}c_1 + c_2) + a_{31}c_1 + c_3 \\ (a_{43}a_{32} + a_{42})(a_{21}c_1 + c_2) + a_{43}(a_{31}c_1 + c_3) + c_4 \end{bmatrix}.$$

The parallel evaluation with $m = 2, n = 4$ requires at most 5 time steps.

Another method for a linear system to be solved in parallel mode is called *parallelization by permutation*. Consider the tridiagonal linear system

$$Ax = d \tag{9.31}$$

with

$$d := [d_1, \ldots, d_n]^T \in \mathbb{R}^n$$

and

$$A := \begin{bmatrix} a_1 & b_1 & & & \\ c_2 & a_2 & b_2 & & 0 \\ & \ddots & \ddots & \ddots & \\ 0 & & \ddots & \ddots & b_{n-1} \\ & & & c_n & a_n \end{bmatrix}.$$

At first sight the SOR method

$$a_1 x_1^{(k+1)} := (1-\omega)a_1 x_1^{(k)} \qquad\qquad -\omega b_1 x_2^{(k)} + \omega d_1$$
$$a_i x_i^{(k+1)} := (1-\omega)a_i x_i^{(k)} - \omega c_i x_{i-1}^{(k+1)} - \omega b_i x_{i+1}^{(k)} + \omega d_i$$
$$a_n x_n^{(k+1)} := (1-\omega)a_n x_n^{(k)} - \omega c_n x_{n-1}^{(k+1)} \qquad\qquad + \omega d_n$$

for $i = 2, \ldots, n-1$, does not seem to allow parallelization. By using the permutation (see also [62])

$$PAP^T y := \bar{A}y = z = \bar{d} \tag{9.32}$$

with
$$y = [x_1, x_3, \ldots, x_{2^k-1}, x_2, \ldots, x_{2^k}]^T$$
$$\bar{d} = [d_1, d_3, \ldots, d_{2^k-1}, d_2, \ldots, d_{2^k}]^T$$
and

$$\bar{A} := \left[\begin{array}{cccc|cccc} a_1 & & & & b_1 & & & \\ & a_3 & & & c_3 & b_3 & & \\ & & \ddots & & & \ddots & \ddots & \\ & & & a_{2^k-1} & & & c_{2^k-1} & b_{2^k-1} \\ \hline c_2 & b_2 & & & a_2 & & & \\ c_4 & b_4 & & & & a_4 & & \\ & \ddots & \ddots & & & & \ddots & \\ & & b_{2^k-2} & & & & & \\ & & c_{2^k} & & & & & a_{2^k} \end{array} \right]$$

the SOR method, applied to $\bar{A}y = \bar{d} = z$, gives:
$$a_i y_i^{(k+1)} = (1-\omega) a_i y_i^{(k)}$$
$$a_{2i-1} y_i^{(k+1)} = (1-\omega) a_{2i-1} y_i^{(k)} - \omega c_{2i-1} y_{\frac{n}{2}+i-1}^{(k)} - \omega b_{2i-1} y_{\frac{n}{2}+i}^{(k)} + \omega z_i$$
$$\text{for } i = 2, \ldots, \frac{n}{2}. \tag{9.33}$$

$$a_{2i} y_{\frac{n}{2}+i}^{(k+1)} = (1-\omega) a_{2i} y_{\frac{n}{2}+i}^{(k)} - \omega c_{2i} y_i^{(k+1)} - \omega b_{2i} y_{i+1}^{(k+1)} + \omega z_{\frac{n}{2}+i}$$
$$a_n y_n^{(k+1)} = (1-\omega) a_n y_n^{(k)} - \omega c_n y_{\frac{n}{2}}^{(k+1)} + \omega z_n$$
$$\text{for } i = 1, 2, \ldots, \frac{n}{2} - 1. \tag{9.34}$$

If $y_1^{(k)}, \ldots, y_n^{(k)}$ are known, then $y_1^{(k+1)}, \ldots, y_{\frac{n}{2}}^{(k+1)}$ are computed in parallel by (9.33) and $y_{\frac{n}{2}+1}^{(k+1)}, \ldots, y_n^{(k+1)}$ are computed in parallel by (9.34)

Parallelization by permutation is the basic concept for the *odd-even reduction method*.

9.6 Parallel solution of linear systems on MIMD machines

To illustrate how a parallel solution method can be designed for an $MIMD$ machine (for example HEP), consider solving a set of symmetric algebraic equations

$$\begin{array}{ccccccc} a_{11} & + & a_{12} x_2 & + & \ldots & + & a_{1n} x_n & = & b_1 \\ a_{21} & + & a_{22} x_2 & + & \ldots & + & a_{2n} x_n & = & b_2 \\ \vdots & & \vdots & & \vdots & & \vdots & & \vdots \\ a_{n1} & + & a_{n2} x_2 & + & \ldots & + & a_{nn} x_n & = & b_n, \end{array}$$

9.6 Parallel solution of linear systems on MIMD machines

which is equivalent to the system

$$Ax = b, \qquad (9.35)$$

with

$$\begin{aligned} A &:= [a_{ik}], \quad i,k = 1,2,\ldots,n \\ b &:= [b_1,\ldots,b_n]^T \\ x &:= [x_1,\ldots,x_n]^T. \end{aligned}$$

For simplicity, it is assumed that row or column interchange for numerical stability is not needed. Also considered is only a factorization of the matrix $A = [a_{ik}]$. A sequential program which produces the lower triangular factor of the matrix A might be as follows:

```
for k := 1 to n − 1 do
  for j := k + 1 to n do
```

$$\left.\begin{aligned} &\textbf{begin} \\ &c := a(j,k)/a(k,k); \\ &\textbf{for } i := j \textbf{ to } n \textbf{ do} \\ &a(i,j) := a(i,j) - a(i,k)*c \\ &\textbf{end}; \end{aligned}\right\} T_j^k$$

The computational task between **begin** and **end** is denoted by T_j^k for given values k and j.

The following Figure 9.11 illustrates the tasks to be executed and their temporal precedence constraints:

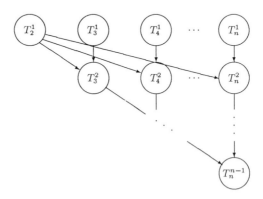

Fig. 9.11 – Executing tasks

It follows, in particular, that the tasks T_3^2 cannot be executed before T_3^1 is completed since both tasks change the values of the third column of A. Also

T_3^2 requires completion of T_2^1 whose elements are used to evaluate components of T_3^2. However, the tasks T_j^2, $j = 3, 4, \ldots, n$, can be computed in parallel once the tasks T_j^1, $j = 2, 3, \ldots, n$, have been finished. It can be shown that even if the number of available processors is high (approximately $n/2$) an efficient parallel algorithm can be constructed. Such an algorithm would keep busy 66% of the parallel processors during the execution period required to decompose the matrix A. If the number of processors p is considerably smaller than n the computational efficiency would quickly increase (above 90% for the HEP). This means that the speed-up provided by the HEP-machine working in parallel mode is roughly $S_p = 8 \cdot r$, where r is the number of *Process Execution Mudules* (*PEM's*).

References

[1] Blumenfeld, M., Doll, J., & Schendel, U., *Numerische Algorithmen zur Lösung von Eigenwertproblemen bei Matrizen*, Brennpunkt Kybernetik, Band 56, Technische Universität Berlin (1974).

[2] Borodin, A., & Munro, I., *The Computational Complexity of Algebraic and Numeric Problems*, American Elsevier (1975).

[3] Brent, R.P., *The parallel evaluation of general arithmetic expressions*, J.ACM, **21**, 2, 201-206 (1974).

[4] Bunch, J.R., & Rose, D.J. (eds.), *Sparse Matrix Computations*, Academic Press, New York (1976).

[5] Businger, P.A., *Monitoring the numerical stability of Gaussian elimination*, Numerische Mathematik, 16 (1971).

[6] Deo, N., *Graph Theory with Application to Engineering and Computer Science*, Prentice Hall, Inc. (1974).

[7] Doll, J., *Ein iteratives Verfahren zur Lösung des nichtsymmetrischen Eigenwertproblems*, Dissertation, Freie Universität Berlin (1980).

[8] Duff, I.S., & Reid, J.K., *A comparison of sparsity orderings for obtaining a pivotal sequence in Gaussian elimination*, J. Inst. Maths. Applics., 14, 281-291 (1974).

[9] Erisman, A., & Reid, J.K., *Monitoring the stability of the triangular factorization of a sparse matrix*, Numerische Mathematik, Band 22 (1974).

[10] Erisman, A., & Tinney, W., *On computing certain elements of the inverse of a sparse matrix*, Comm. ACM, **10**, 3 (1975).

[11] Evans, D.J., *Software for Numerical Mathematics*, Proceedings of the Loughborough University of Technology, Academic Press, London, New York (1974).

[12] Evans, D.J., *Iterative sparse matrix algorithms. In Evans, D.J. (ed.), Software for Numerical Mathematics*, Proceedings of the Loughborough University of Technology, Academic Press, London New York (1974).

[13] Feilmeier, M., Joubert, G., & Schendel, U. (eds), *Parallel Computing 83*, North-Holland, Amsterdam (1984).

[14] Feilmeier, M., Joubert, G., & Schendel, U. (eds), *Parallel Computing 85*, North-Holland, Amsterdam (1986).

[15] Fiedler, M., *A property of eigenvectors of nonnegative symmetric matrices and its application to graph theory*, Czechoslovak Mathematical Journal, No. 4 (1975).

[16] Flynn, M.J., *Very high speed computing systems*, Proc. IEEE, **14**, 1901-1909 (1966).

[17] George, J.A., *On block elimination for sparse linear systems*, SIAM J. Numer. Anal., **11**, 585-603 (1974).

[18] Giloi, W.K., *Rechnerarchitektur*, Springer-Verlag Berlin, Heidelberg, New York (1981).

[19] Golm, K., & Schendel, U., *On special parallel algorithms for parabolic differential equations*, Parallel Computing 85, North-Holland, Amsterdam (1986).

[20] Golub, G.H., & van Loan, C.F., *Matrix Computations*, The Johns Hopkins University Press, Baltimore, Maryland (1983).

[21] Harary, F., *Graph Theory*, Academic Press (1969).

[22] Hockney, R.W., & Jesshope, C.R., *Parallel Computers*, Adam Hilger Limited (1981).

[23] Hwang, K., *Tutorial Supercomputers: Design and Applications*, IEEE Computer Society Press, Silver Spring, MD (1984).

[24] Hyafil, L., & Kung, H.T., *The complexity of parallel evaluation on linear recurrences*, J.ACM, **24**, 3, 513-521 (1974).

References

[25] James, K.R., & Riha, W., *Convergence criteria for successive overrelaxation*, SIAM J. Numer. Anal., **12**, 2 (1975).

[26] Kahan, W., *Gauss-Seidel methods of solving large systems of linear equations*, Doctoral Thesis, University of Toronto, Torònto, Canada (1958).

[27] Kogge, P.M., *Parallel solutions of recurrence problems*, IBM J. Res. Develop., **18**, 138-148 (1974).

[28] Kowalik, J.S. (ed.), *Parallel MIMD Computation: HEP Supercomputers and its Application*, The MIT Press, Cambridge, Massachusetts (1985).

[29] Kuck, D.J., *Structures of Computers and Computations*, John Wiley & Sons, New York (1978).

[30] Lancaster, P., *Lambda-Matrices and Vibrating Systems*, Pergamon, Oxford (1966).

[31] Liniger, W.M., & Willoughby, R.A., *Efficient numerical integration of stiff systems of ordinary differential equations*, SIAM J. Numer. Anal., **7**, 47-66 (1970).

[32] Mead, C.A., & Conway, L.A., *Introduction to VLSI systems*, Addison-Wesley, Reading, Mass. (1980)

[33] Metropolis, N., Sharp, D.H., Worlton, W.J., & Ames, K.R. (eds), *Frontiers of Supercomputing*, University of California Press, Berkeley (1986).

[34] Moto-Oka, T. (ed.), *Fifth Generation Computer Systems*, North-Holland, Amsterdam (1982).

[35] Müller-Wichards, D., & Gentzsch, W., *Performance comparisons among several parallel and vector computers on a set of fluid flow problems*, DVLR, Report No. IB262-82RO1, Göttingen (1982).

[36] Ostrowski, A.M., *On the linear iteration procedures for symmetric matrices*, Rend. Mat. Appl., **14**, 140-163 (1954).

[37] Paton, K., *An algorithm for the blocks and cutnodes of a graph*, Comm. ACM, **13**, 7, 446-449 (1967).

[38] Read, R., *Graph Theory and Computing*, Academic Press, New York (1972).

[39] Reid, J.K., *Large Sparse Sets of Linear Equations*, Academic Press, New York (1971).

[40] Rogers, L.D., *Optimal Paging Strategies and Stability Considerations for Solving Large Linear Systems*, Dissertation, University of Waterloo, Ontario (1973).

[41] Rose, D.J., & Willoughby, R.A., *Sparse Matrices and their Applications*, Plenum Press, New York (1972).

[42] Ruhe, A., *Iterative Eigenvalue Algorithms for Large Symmetric Matrices*, Report No. 31-72, Umea University, November (1973).

[43] Rutishauser, H., *Simultaneous iteration method for symmetric matrices*. In Wilkinson, J.H., & Reinsch, C., *Handbook for Automatic Computation*, Linear Algebra, vol.2, Springer Verlag, New York (1971).

[44] Schendel, U., *Sparse-Matrizen und Anwendungen*, Brennpunkt Kybernetik, Technische Universität Berlin, Band 60 (1975).

[45] Schendel, U., *Graphentheoretische Methode zur Blockstrukturierung symmetrischer Matrizen*, Preprint No. 21/1977, Freie Universität Berlin (1977).

[46] Schendel, U., *On sparse matrix computation*, Report No. CS.07.03.06.78, Comp. Sc. Dept., Univ. of Natal, Durban (1978).

[47] Schendel, U., & Brandenburger, J., *Algorithmen zur Lösung rekurrenter Relationen*, Preprint No. 101/79, Freie Universität Berlin (1979).

[48] Schendel, U., *On basic concepts in parallel numerical mathematics*, CONPAR 81, Springer-Verlag Berlin, Heidelberg, New York (1981).

[49] Schendel, U., *Introduction to Numerical Methods for Parallel Computers*, Ellis Horwood Limited, Chichester (1984).

[50] Schendel, U., & Westphal, K.- P., *MIMD-Rechner und Anwendungen*, Preprint No. 203/85, Freie Universität Berlin (1985).

[51] Schwarz, H.R., *The method of coordinate overrelaxation for $(A-\lambda B)x=0$*, Numerische Mathematik, 23 (1974).

[52] Seppänen, J.J., *Algorithm 399: spanning tree*, Comm. ACM, **13**, 10, 621-622 (1972).

[53] Stein, P., & Rosenberg, R., *On the solution of linear simultaneous equations by iteration*, J. London Math. Soc., **23**, 111-118 (1948).

[54] Stewart, G.W., *Introduction to Matrix Computation*, Academic Press, New York (1973).

[55] Strang, G., & Fix, G.J., *An Analysis of the Finite Element Method*, Prentice Hall, Englewood Cliffs, New York (1973).

[56] Takahashi, K., Fagan, J., & Chen, M., *Formation of a sparse bus impedance matrix and its application to short circuit study*, 8th PICA Conference Proceedings, 4-6 June (1973).

[57] Tewarson, R., *Sparse Matrices*, Academic Press, New York (1973).

[58] Wilkinson, J.H., *Rounding Errors in Algebraic Processes*, Her Majesty's Stationery Office, London, Prentice-Hall, New Jersey (1963).

[59] Wilkinson, J.H., *The Algebraic Eigenvalue Problem*, Oxford University Press, London (1965).

[60] Wilkinson, J.H., & Reinsch, C., *Handbook for Automatic Computation*, Linear Algebra, vol.2, Springer Verlag, New York (1971).

[61] Wouk, A. (ed.), *New Computing Environments: Parallel, Vector and Systolic*, SIAM, Philadelphia (1986).

[62] Young, D.M., *Iterative Solution of Large Linear Systems*, Academic Press, New York (1971).

Index

adaptive methods, 3, 88
arithmetical expression, 92, 95
    equivalent, 94
    minimal evaluation of, 93
arrow, 73, 74
    composition, 76
    of reverse direction, 75
    parallel, 73, 74

backward substitution, 30, 31, 35, 44
band width, 12, 13, 39, 79
block diagonalization, 78
block iteration, 64

characteristic polynomial, 80
condition number, 27, 48
condition,
    of a matrix, 40, 42
convergence,
    of coordinate relaxation, 85
    of Gauss-Seidel method, 57-61, 66, 68
    of iterative algorithms, 47

    of Jacobi method, 54-58, 60, 68
    of power iteration, 86
    of QR-algorithm, 82
    rate of, 48
    of relaxation method, 83
    of SOR method, 61-63, 66-70
coordinate overrelaxation, 83
    algorithm of, 84
coordinate relaxation, 83-85

dag, 93
decomposition,
    block, 19, 30, 38
    Cholesky, 37, 38, 40, 42, 65
    LU-, 23, 32, 34, 40, 65
    QR-, 43, 44
    triangular, 13, 19, 28
differential equation,
    ordinary, 1
    partial, 1, 40, 69, 91
        self-adjoint, 3
        self-adjoint elliptic, 2, 64

Index

edge, 73
eigenvalue, 1, 28, 47, 51, 54, 56,
    58-63, 67-69, 71, 78
  all of a matrix, 82
  several, 86
  smallest, 82, 83, 85
  strictly dominant, 86
eigenvalue problem, 1, 78
  conventional, 80
  general, 79, 82-84
  nonsymmetric, 80
  symmetric, 83
eigenvector, 67-70
  approximate, 85
error equations, 41, 44
  normal form, 42
extrapolation, 61
extrapolation techniques, 47

factorization, 20, 25, 28, 30, 31, 105
  block, 30
  LU-, 5, 25, 40
feed-back-node problems, 93
fill-in, 5, 22, 35, 47, 77, 81
  block, 30
  local, 20-22
finite differences, 28, 40, 79
finite elements, 4, 28, 40, 79

Gauss decomposition, 36
Gauss transformation, 42
Gaussian elimination, 13, 14, 20-23,
    28-30, 34, 39, 77, 91
Gauss-Seidel method, 57-62, 66, 68
Gauss-Seidel value, 61
graph, 72, 73, 95
  acyclic, 93
  complete, 62
  connected, 73, 77
  digraph, 73-77
    asymmetrical, 75
    complete, 76
    quasi-strictly connected, 78
    representation by matrices, 75
    strictly connected, 77, 78
    symmetric, 75

transitive, 76
directed, 73
equal, 74
loop of, 73, 74
graph theory, 72, 77

Hessenberg form, 80, 81
Horner algorithm, 94
Householder transformation, 44, 80,
    81

item, 6, 8
iterated vectors, 85, 86

Jacobi method, 53, 55-58, 60, 61, 68

Krylow-sequences, 85

matrix,
  achievement, 78
  adjacency, 26, 72, 76-78
  band, 2, 12, 13, 39, 40, 79, 98
    circular, 13
    with margin, 3
    with step, 3
  block, 29, 37, 38, 64, 66
    diagonal, 3, 36, 78
    inverse, 37
    partitioning, 36
    tridiagonal, 36, 38, 66
  decomposable, 18, 29, 48
  defective, 80
  diagonally dominant, 12, 70, 71
    generalized, 66-68, 70
    strictly, 66, 67, 69
    weakly, 28, 48, 50-52, 56, 58,
      62, 66, 69
  Frobenius, 15
  Hessenberg, 81, 82
  inverse, 17, 27, 34, 35, 39, 65, 78
    elements of the inverse, 25, 26
  irreducible, 29, 34, 48-52, 56, 58,
    59, 62, 66-70, 77, 78
  Jacobi, 12
  L-matrix, 52, 56, 60, 66, 68
  M-matrix, 60

non-decomposable, 48
N-stable, 71
permutation, 20, 29, 32, 35, 36, 52, 77, 78
reducible, 29, 48, 49
sparse, 2
  inverse, 25
  structure of, 2, 3, 22, 47
strip, 3
symmetric, 23, 26
triangular, 14, 15, 17-20, 23, 25, 26, 28, 31-35, 45, 58, 105
  quasi right-hand, 82
  regular, 43
  right-hand, 82
  strictly, 58, 60, 67, 69, 98, 101
tridiagonal, 39, 80, 103
  symmetric, 82
  unitary, 82
method of least squares, 41
minimization problem, 42

Newton method, 12
node-parse-complete problem, 92

orthogonal transformation, 42
overrelaxation, 61, 83
paging strategies, 6
parallel algorithm, 89
  applications of, 91
  characterization of, 90
  column-sweep algorithm, 101
  MIMD algorithm, 90, 91
  parallelization by permutation, 103, 104
  product form algorithm, 101
  recursive doubling, 96, 102
  SIMD algorithm, 90, 91
  SOR method, 103, 104
  systolic algorithm, 90, 91
parallel computer, 87, 92
  array machine, 91, 92
  characteristics of, 88
  general model, 89
  MIMD machine, 88, 89, 95, 97, 104

MISD machine, 88
SIMD machine, 88, 89, 95, 96, 99
SISD machine, 88
parsetree, 97, 100
Perron-Frobenius theory, 67
pivot, 20, 39, 40, 81
  element, 13-15, 20-22, 28, 40
  tolerance, 21, 22
power iteration, 80, 86
  shifted, 86
propagation property, 31
property A, 52
property P, 31-35

QR-algorithm, 82
QR-transformation, 82

rate of convergence,
  of Gauss-Seidel method, 61
  of Jacobi method, 61
  of SOR method, 70
Rayleigh quotient, 82-85
recurrence relation, 98, 102
  column-sweep algorithm, 101
  companion function, 99, 100
  first order, 98
  function, 99
  general, 99, 100
  linear, 98, 101
  ordinary, 98, 101
  product form algorithm, 101
  recursive doubling, 102
  second order, 98
relaxation factor, 61, 84
relaxation method, 83
residual vector, 41, 42, 44
Richardson method, 48

simultaneous displacement method, 48
simultaneous iteration, 86
singular value, 45, 46
singular value decomposition, 45
singular vector, 45, 46

SOR method, 61, 63-66, 68, 69
    block, 65
    parallel, 103, 104
spectral radius,
    of Gauss-Seidel method, 60
    of Jacobi method, 54, 55
    of SOR method, 67, 69
stability, 22, 24, 36
storage techniques, 5, 8, 36, 47
successive overrelaxation method, 61
system of linear equations, 1, 6, 13,
        14, 22, 36, 38, 79, 81, 88,
        103
    overdetermined, 41

tree height reduction, 95, 97
triangulation, 4, 18, 19, 35
    block, 39, 78

vertex, 73, 74, 76
    final, 74
    initial, 74

weak row sum criterion, 48

Mathematics and its Applications

Series Editor: G. M. BELL, Professor of Mathematics, King's College London (KQC), University of London

Gardiner, C.F.	Algebraic Structures
Gasson, P.C.	Geometry of Spatial Forms
Goodbody, A.M.	Cartesian Tensors
Goult, R.J.	Applied Linear Algebra
Graham, A.	Kronecker Products and Matrix Calculus: with Applications
Graham, A.	Matrix Theory and Applications for Engineers and Mathematicians
Graham, A.	Nonnegative Matrices and Applicable Topics in Linear Algebra
Griffel, D.H.	Applied Functional Analysis
Griffel, D.H.	Linear Algebra and its Applications: Vol 1, A First Course; Vol. 2, More Advanced
Guest, P. B.	The Laplace Transform and Applications
Hanyga, A.	Mathematical Theory of Non-linear Elasticity
Harris, D.J.	Mathematics for Business, Management and Economics
Hart, D. & Croft, A.	Modelling with Projectiles
Hoskins, R.F.	Generalised Functions
Hoskins, R.F.	Standard and Non-standard Analysis
Hunter, S.C.	Mechanics of Continuous Media, 2nd (Revised) Edition
Huntley, I. & Johnson, R.M.	Linear and Nonlinear Differential Equations
Irons, B. M. & Shrive, N. G.	Numerical Methods in Engineering and Applied Science
Ivanov, L. L.	Algebraic Recursion Theory
Johnson, R.M.	Theory and Applications of Linear Differential and Difference Equations
Johnson, R.M.	Calculus: Theory and Applications in Technology and the Physical and Life Sciences
Jones, R.H. & Steele, N.C.	Mathematics in Communication Theory
Jordan, D.	Geometric Topology
Kelly, J.C.	Abstract Algebra
Kim, K.H. & Roush, F.W.	Applied Abstract Algebra
Kim, K.H. & Roush, F.W.	Team Theory
Kosinski, W.	Field Singularities and Wave Analysis in Continuum Mechanics
Krishnamurthy, V.	Combinatorics: Theory and Applications
Lindfield, G. & Penny, J.E.T.	Microcomputers in Numerical Analysis
Livesley, K.	Engineering Mathematics
Lootsma, F.	Operational Research in Long Term Planning
Lord, E.A. & Wilson, C.B.	The Mathematical Description of Shape and Form
Malik, M., Riznichenko, G.Y. & Rubin, A.B.	Biological Electron Transport Processes and their Computer Simulation
Massey, B.S.	Measures in Science and Engineering
Meek, B.L. & Fairthorne, S.	Using Computers
Menell, A. & Bazin, M.	Mathematics for the Biosciences
Mikolas, M.	Real Functions and Orthogonal Series
Moore, R.	Computational Functional Analysis
Murphy, J.A., Ridout, D. & McShane, B.	Numerical Analysis, Algorithms and Computation
Nonweiler, T.R.F.	Computational Mathematics: An Introduction to Numerical Approximation
Ogden, R.W.	Non-linear Elastic Deformations
Oldknow, A.	Microcomputers in Geometry
Oldknow, A. & Smith, D.	Learning Mathematics with Micros
O'Neill, M.E. & Chorlton, F.	Ideal and Incompressible Fluid Dynamics
O'Neill, M.E. & Chorlton, F.	Viscous and Compressible Fluid Dynamics
Page, S. G.	Mathematics: A Second Start
Prior, D. & Moscardini, A.O.	Model Formulation Analysis
Rankin, R.A.	Modular Forms
Scorer, R.S.	Environmental Aerodynamics
Shivamoggi, B.K.	Stability of Parallel Gas Flows
Smith, D.K.	Network Optimisation Practice: A Computational Guide
Srivastava, H.M. & Manocha, L.	A Treatise on Generating Functions
Stirling, D.S.G.	Mathematical Analysis
Sweet, M.V.	Algebra, Geometry and Trigonometry in Science, Engineering and Mathematics
Temperley, H.N.V.	Graph Theory and Applications
Temperley, H.N.V.	Liquids and Their Properties
Thom, R.	Mathematical Models of Morphogenesis
Thurston, E.A.	Techniques of Primary Mathematics
Toth, G.	Harmonic and Minimal Maps and Applications in Geometry and Physics
Townend, M. S.	Mathematics in Sport
Townend, M.S. & Pountney, D.C.	Computer-aided Engineering Mathematics
Trinajstic, N.	Mathematical and Computational Concepts in Chemistry
Twizell, E.H.	Computational Methods for Partial Differential Equations
Twizell, E.H.	Numerical Methods, with Applications in the Biomedical Sciences
Vince, A. and Morris, C.	Mathematics for Information Technology
Walton, K., Marshall, J., Gorecki, H. & Korytowski, A.	Control Theory for Time Delay Systems
Warren, M.D.	Flow Modelling in Industrial Processes
Wheeler, R.F.	Rethinking Mathematical Concepts
Willmore, T.J.	Total Curvature in Riemannian Geometry
Willmore, T.J. & Hitchin, N.	Global Riemannian Geometry

Numerical Analysis, Statistics and Operational Research
Editor: B. W. CONOLLY, Emeritus Professor of Mathematics (Operational Research), Queen Mary College, University of London

Beaumont, G.P.	Introductory Applied Probability
Beaumont, G.P.	Probability and Random Variables
Conolly, B.W.	Techniques in Operational Research: Vol. 1, Queueing Systems
Conolly, B.W.	Techniques in Operational Research: Vol. 2, Models, Search, Randomization
Conolly, B.W.	Lecture Notes in Queueing Systems
Conolly, B.W. & Pierce, J.G.	Information Mechanics: Transformation of Information in Management, Command, Control and Communication
French, S.	Sequencing and Scheduling: Mathematics of the Job Shop
French, S.	Decision Theory: An Introduction to the Mathematics of Rationality
Griffiths, P. & Hill, I.D.	Applied Statistics Algorithms
Hartley, R.	Linear and Non-linear Programming
Jolliffe, F.R.	Survey Design and Analysis
Jones, A.J.	Game Theory
Kapadia, R. & Andersson, G.	Statistics Explained: Basic Concepts and Methods
Moscardini, A.O. & Robson, E.H.	Mathematical Modelling for Information Technology
Moshier, S.	Mathematical Functions for Computers
Oliveira-Pinto, F.	Simulation Concepts in Mathematical Modelling
Ratschek, J. & Rokne, J.	New Computer Methods for Global Optimization
Schendel, U.	Introduction to Numerical Methods for Parallel Computers
Schendel, U.	Sparse Matrices
Schmidt, N.S.	Large Order Structural Eigenanalysis Techniques: Algorithms for Finite Element Systems
Späth, H.	Mathematical Software for Linear Regression
Spedicato, E. and Abaffy, J.	ABS Projection Algorithms
Stoodley, K.D.C.	Applied and Computational Statistics: A First Course
Stoodley, K.D.C., Lewis, T. & Stainton, C.L.S.	Applied Statistical Techniques
Thomas, L.C.	Games, Theory and Applications
Whitehead, J.R.	The Design and Analysis of Sequential Clinical Trials